The Ess Wild Food Survival Guide

Copyright © 2002, 2007 by Linda Runyon
original title *From Crabgrass Muffins to Pine Needle Tea*

Linda Runyon
Wild Food Company
PO Box 83
Shiloh, NJ 08353-0083
lrunyon8@yahoo.com
www.OfTheField.com

ISBN 0-936699-10-8

All Rights Reserved, including the right to reproduce this book,
or parts thereof, in any form, except for the
inclusion of brief quotations in a review.

All Art by Linda Runyon unless otherwise noted

Photo Credits:
Ruth Demers Executive Director of Rome Sports Hall of Fame
of Rome, NY – all photos except the following:
Kerin A. Denser-Fuzy – sow thistle
Paul A. Runyon, Jr. of Seattle, WA – prickly pear, saguaro,
cholla, lamb's quarters, malva neglecta, sunflower
& shepherd's purse.

Main Cover Art by Linda Runyon
Cover Design by Eric & Todd Conover
Layout and Final Production by Eric Conover

Disclaimer

This book is intended to be an educational tool for gathering and cooking wild plants. The information presented is for use as a supplement to a healthy, well-rounded lifestyle. The nutritional requirements of individuals may vary greatly, therefore the author and publisher take no responsibility for an individual using and ingesting wild plants.

Steps with lamb's quarter seedlings.

Waste ground plantain dandelion.

Dedication

I dedicate this manuscript to Ruth and Paul Runyon who gave me the feeling of unconditional love. Thank you, Mom and Dad. You made it all happen!

Thank you, my beautiful children whose understanding of Mom kept me on the "wild" path. Without my son Eric it would not be possible to bring you this revised edition. Over the years the help has been invaluable. My grandchildren Ceri & Alec, Jason, and Lee, who have been my guinea pigs with spearmint gum and dandelion fluff popsicles.

The homestead years were wonderful and difficult. Thanks to Ken Heitz, who spent them supporting us and our homestead! Deepest thanks to the late Dorothy and Lynn Galusha, Gert and Dick Frulla, Indian Lake, and dozens of friends: especially Nancy & Pete, Nicole, Colleen & Jenny, Jean, Terry, Shirley and Eric & Rose. I couldn't have done it without you. Thanks to Gary for computer help.

A special acknowledgment to Cristina, one of my teachers, and her mom, Melinda. Out of the woods, the Runyon Institute began. Thank you, Walt Johnson, you taught me the business world! My deepest gratitude to Mae Charon, without whose friendship and business guidance I never would have completed this manuscript. Bruce & Larry, your faith, love and encouragement are monumental! Alice & the late Walt Sherman, your love is my port in the storm. This book was conceived in the lamb's quarter patch behind the barn. Ruth Demers, you crawled on your tummy, festooned with black flies to take these photos, thank you. Your faith became mine. Paul, my "bro", you stood supporting me in 110 degrees while you took the western slides! Thank you!

So many people, thank you all!!! Thanks especially Dr. Edward Glenn and the Environmental Research Lab in Tucson for your library, approval and guidance along the path to the Environmentarian Way. Susan Lake and Nancy Sprague gave me their invaluable library skills. Steve "Wildman" Brill is the foraging cohort I trust implicitly. Your advice is well taken. Joan & David Schultz, Kathy & Mel, Richard & Stephanie Bond, Tess and Michael Morearty, you all give me a tremendous feeling of support here in Arizona.

The Wild Food Walks were the result of gracious offers for the use of private lands. Thank you Rose and Bruce Burke, the Nanticote Lenni Lenape tribe of South Jersey, Jim Ridgeway; Kathy Weaver Warrensburg; Lenore Yalisloff and Hazel McManus in Phoenix, Glendale Public Library and David Schultz's direction here in Arizona.

Thanks to Dr. James Duke, Ph.D. you are the database for my manuscript! Steve "Wildman" Brill edited professional to professional. Althea Dixon, your generosity kept my newsletter going...Wanda Wilde, secretary for Wild Foods Co., thank you for your help and hello to your 6 kids (including the triplets)! I know there is a list of hundreds over the years—you all know who you are!!! Thank you! An abundant green year awaits us all.

Miss Linda Runyon

Todd's special place - the cave!

Table of Contents

	Page No.
Dedication	i
Foreword	iv
Preface	vi
Acknowledgements	viii

PART I AN ENVIRONMENTARIAN LIFESTYLE
Living off the Land	1
An Introduction to Edible Wild Plants	17

PART II FIELD GUIDE TO WILD FOODS
Edible Plants	49
Photos of the Plants	141
Nutrient Value of Wild Foods	173

PART III WILD FOOD RECIPES — 195
Soups and Salads	198
Vegetable Dishes & Light Meals	205
Muffins, Breads, Biscuits, and Pancakes	225
Sweets	233
Teas and Other Hot and Cold Drinks	240
Preserving Wild Foods	250
Sundries	255

PART IV REFERENCE SECTION
Cultivation of Wild Edibles	259
Public Wild Food ID Walk Ways	270
A Forager's Research Diary	283

PART V POISONOUS LOOK-ALIKES — 290

Appendix Glossary of Scientific & Botanical Terms	302
References	305
Suggested Reading	306

Foreword —by Joe Gang

Linda Runyon is a child of the Adirondack wilderness. It is there that, at a tender age, she discovered that wild clover, yarrow, and wintergreen were edible. As she grew, so grew her knowledge of the abundant wild flora in her beloved mountains.

In 1972 she decided to homestead in the wilderness. Her ability to recognize and use wild plants added immeasurably to her successful survival. By adapting to a diet of wild vegetables, herbs, fruits, and nuts, Linda carved out a niche for herself among women pioneers who relied on Nature. Linda Runyon has gone on to share her wide-ranging experiences and knowledge of wild foods with a growing following: visitors to several plant identification walks (called *Survival Acres*), readers of her books, (*A Survival Acre, Lawn Food Cookbook,* and *Wild Food and Animal Coloring Book*), users of her fantastic plant identification card game ("Wild Cards"), and subscribers to her (now out of print) Environmentarian Newsletter, and the "Wild Foods Nutritional Wheel", added later. Linda Runyon's valuable *From Crabgrass Muffins to Pine Needle Tea* opened a vast fund of knowledge on nature's own foodstuffs, which often can be found close to home, in an environment unsoiled by insecticides, herbicides, or fungicides. Here is the key to the preparation of wild foods without the use of preservatives, invasive chemicals or factory processing.

And now we present a new edition of that book renamed to *The Essential Wild Food Survival Guide,* with all the sustaining data from the first edition. This book will continue to be your valuable companion on the path to healthful living.

The author offers a new way to view the world.

Poem

MY LAWN
As I walked across the grass,
I noticed clover, dandelion, plus
a few plantains.
The thought came to me that
I should graze on all four.
All other modes of feeding
now seem so inefficient,
even wasteful.
When I go out to graze,
I'll carry a bottle of salad dressing,
and sprinkle liberally in front of me,
as I munch my way across the lawn.

By Joe Gang Environmentalist, master of Economics and Finances, chemist, small business consultant, Indian Lake, New York.

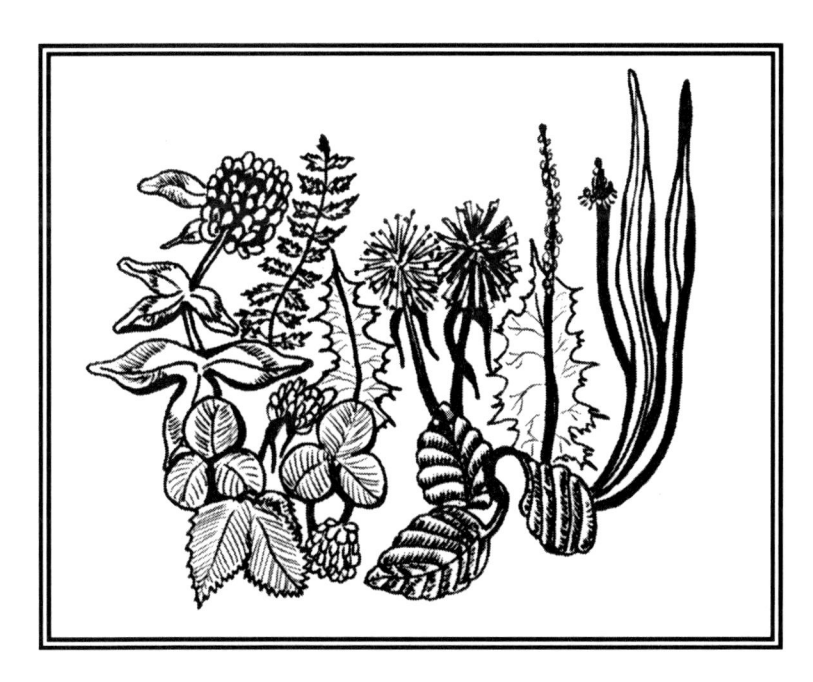

A typical Eastern lawn smorgasbord

Preface

I am called an environmentarian, or a forager in common terms. I enjoy foraging in fields, woods, and even backyards for the wild plants that make up as much as 30 percent of my daily diet.

The knowledge that led me to this way of life did not come overnight. Walking eight and a half miles to work each day one summer taught me the value of knowing the energy in a single clover leaf; after 15 years of homesteading, (living without any power or electricity in cabins and tents) I learned the tricks of gathering and cooking nature's food. It is quick growing, accessible, nutritious, and free!

Many years of this experience taught me not only how to forage, what to eat, and how to prepare it, but it has also confirmed my decision to stay in touch with my primal rights to live a free life. Now that I'm over 70 years old, my goal is to share this knowledge with others from a comfortable retirement area.

I have divided this book into four major parts, and I've also included some backup information in a resource section and glossary. Part I describes my environmentarian lifestyle, including how I came to adopt this way of life and how I learned and acquired the wilderness skills that helped me survive. In this section, I also provide information on gathering wild plants as well as how to store them for later use—drying, freezing, and the like.

Because properly identifying wild foods in the field is critical, I have provided a small Field Guide in Part II, which pictures and describes over 50 wild plants, including grasses, herbs, brambles, and trees, plus their nutritional values.

In Part III, I provide more recipes that use wild foods as the major ingredients—tasty and easy-to-prepare soups, salads, casseroles, breads and crackers, sweets, teas, jams & jellies.

In Part IV, I offer tips for growing a wild food garden and for creating "wilderness" walks. I also explain how to grow certain varieties as potted plants, should you live without space for a garden or want to bring your plants indoors. This section also contains my foraging research notes, which tell how I investigated the edibility of wild plants.

Part V discusses Poisonous Look-Alikes and the Glossary defines the terms—mostly medical and botanical—that I use in this book. The book concludes with a listing of useful sources for additional information on wild foods.

It is my hope that this book will help you begin to seek out your own free life, should you wish to do so.

Remember, the process of living free does not happen overnight. In fact, the changes in my lifestyle came very slowly, with gradual adaptation. Try it yourself. If you cannot make it one way, adapt. If you cannot find the food you need, change. If you cannot find shelter, make it. If you are cold, build a fire. If you cannot take a bath, adjust! Such adaptations brought about the changes necessary for me to adopt a free lifestyle.

These changes meant being free of monetary need, not free from constant building, splitting wood, and food seeking. Meeting these needs took up a 24-hour period, with occasional days or hours in between to enjoy the lifestyle. As with any endeavor, you need to enjoy the project. I love projects—they fulfill my day. I always set a goal: move a cord of wood, trim that maple tree, re-screen the windows where the raccoon came in, bathe the mule, clean out the stalls, milk the goats, gather some food for dinner, can the tomatoes.

I plan each day and finish the job when I reached my goals. In between, I feel "on vacation" while I gather food for wild food salads, soups, or casseroles. And, as tired as I am, I have enjoyed every minute of every day.

Linda's Surprise Encounter with a Black Bear
(See page 118)

Acknowledgments

Because of my geographic locations, I use Cherokee and Algonquin language as well as Chippewa throughout the text. For Cherokee translations I have used the CHEROKEE-ENGLISH DICTIONARY, Cherokee Nation of Oklahoma, 1975, Commemorative Edition, edited by William Pulte, Southern Methodist University Press, Dallas, Texas. The dictionary was graciously lent to me by my friend, Joseph Tenderfoot Hill, an Iroquois Elder. I am also privileged to have worked with my friend, a mixed-blood Dijalaga (Cherokee) teacher, Willy Whitefeather.

A Typical Western Smorgasbord

Part I: Environmental Lifestyle

Living off the Land

It seems that the Adirondacks Mountains and their way of life have been a part of my life for as long as I can remember. For this, I have deep gratitude for my parents, Ruth and Paul Runyon. While growing up, I worked on the girl's staff of Gavetts Camp on Indian Lake in upstate NY, now called Timberlock. From the time I was ten years old until I was eighteen, I ran between cabins and tents, carrying army blankets, making beds, and serving 100 guests three meals a day. The dishes and campfire sing-along became a nightly routine for three months of every year. Six A.M. lake dips, canoe races, island campfires, and mountain climbing all led the way to my eventual homesteading experiences. Each year, the return to school in North Plainfield, New Jersey became more and more difficult. Smelling a souvenir balsam pillow linked the mountains to the city for me, and even the taste of wintergreen became a symbol of freedom.

My high school years flew by quickly, and soon I had attended and graduated from Monmouth Medical Center in Long Branch, New Jersey, as a registered nurse. Not long after I married, the first of my family arrived: Eric taught me the wonders of motherhood. I don't have to tell anyone how close your first-born is to your heart. Then within two years, a daughter arrived—Kim, my beautiful daughter, heralded a different bonding in my life. A few years later a second son, Todd, was born. He was to share my entire homesteading experience.

We spent 6 months in a tent.

For many years, all my family thrived in a large housing development in Howell Township, New Jersey. Eventually, my husband Frank and I divorced, and a year or so later I met Ken Heitz. Todd, Ken, and I left New Jersey with an ax, sleeping bags, and great anticipation for the Adirondack wilderness. Kim and Eric remained with their father, but would join us in the summers. Thus my homesteading experience, wild foods, winter survival, and just plain Adirondack life began.

It wasn't long before ideas of an environmentarian lifestyle became a reality. I began to develop the strong bond I feel with Mother Earth. My life was plants, water, shelter, earth, life—"the basics," as someone once said.

Instinct and Ingenuity

One day, in the middle of the winter, I was taken to see a cabin for rent across Lewey Lake. Peering through the cabin's frosty windows, I took less than a minute to shout, "Yes, we'll take it!" An unwinterized cabin, isolated from other summer cabins and framed in black against the mountains, looked like heaven to me. At twenty below, I moved in with Ken and my son Todd, who was age five. We bobsledded the furniture down a moonlit icy path and over twelve-foot snowdrifts to our new home.

Log cabin. Note water hole in frozen lake.

My very first job was to make a pot of coffee. Ken gave me an ax and a pot, and shoveled a path to the frozen lake. Two hours later, time seemed frozen. A two-foot-wide hole grew before me. As I broke through the ice, I felt as if it was ten thousand years ago. Just as folks at the dawn of time had to learn, I, too, soon discovered my ignorance. When the vacuum was released through the newly dug hole, the water rushed up and quickly disappeared below the ice, dragging everything with it. I dove for the coffee pot, only my arm was not long enough to reach deep enough into the water. With tears streaming down my cheeks, I realized my great need for a friend, a tool, and better instincts! Ken provided the first two, and my own instinct led me to my idea for a rope dipper. To this day I can find no words to describe that first cup of coffee!

Water hole tap

Years later, my friend Willie Whitefeather would show me how to gather water in the desert, at the base of a cottonwood tree. Perhaps his Cherokee words, "adline" or "draw liquid," describe that coffee.

Due to a series of events, I learned to respect the Adirondack winter. The first incident occurred when I opened my eyes after a night's sleep. My head seemed stuck to the pillow—my hair had frozen to the frost on the cabin wall! Frosted walls are a feature of life in the far north. A layer of ice 1/4 inch thick covers the walls most of the winter, brought about by intense cold, uninsulated buildings and potbelly heat. My side of the bed was against the wall, and that morning I was literally stuck to the wallboards. Warm water might have solved the problem, but one doesn't always think clearly in a panic! Instead, a haircut solved the problem.

Another morning I awoke and felt a layer of frost on my upper lip and under my nose. I saw my breath as white as smoke, and I knew the stove had gone out during the night. I put my slipper socks over my thermal underwear and hopped to the stove, stirred the coals, heaved in a few pieces of beech, and dove back under the covers. Hearing the crackle of the wood as it flames became my security signal on cold winter mornings.

As I began to learn more about winter, I decided to can my wild foods during the summer for use later in the year. We had a seven-and-a-half-foot insulated refrigerator pit where they would keep well, according to a book I read. I trusted my judgment on this, for I had canned many times in my "city life." So after a long canning season I had 420 jars of campfire-canned foods in my storehouse.

When winter came, we were thrilled to eat our canned foods—cattails, milkweed, lamb's quarters, blueberries, and strawberries. Everything was delicious. Then three weeks of twenty to forty-below zero temperatures began. One night a blizzard snowed us in, but I knew we had plenty of food. The winds howled, the snow swirled, and when morning came, the refrigerator pit had to be dug out. When the roof was uncovered, I opened the pit and climbed down the ladder with my lantern in hand. My eyes blinked in disbelief. Almost every one of the jars had exploded. I was in a complete panic. I grabbed the snow shovel and began to scoop the jars off the shelves to find ones that had just begun to freeze. Alas, there were but a few jars with small cracks or pushed-up lids.

Reconstituted dried leaves, vegetables and flour became our immediate food sources. We even ground bark into flour or brewed it for teas. Somehow, we survived. As I look back, I would say that it was this experience that led me to write this book.

Deer are My Teachers

The first time I looked for food in the woods, I took time to become attuned to the deer's world. I followed deer tracks and saw where and what they had eaten. I found out where they appeared to move quickly, where they moved slowly, and where they stood and relaxed. And when they relaxed, I discovered they were almost always eating.

Well fed fawn

Although not every food eaten by deer is safe for humans, much is.

It became a simple matter to find holes deer had pawed in the snow. These snow diggings yielded wintergreen, ferns, and mosses. I noticed the deer bit off the tips of balsam fir needles. Dried goldenrod tops jutting above the snow line were also nibbled.

Dried-up apples and rosehips also helped to sustain the deer population in the dead of winter. In early spring, the new growth on blackberries, raspberries, willow, and birch added to the deer's diet.

I loved to find an apple tree in winter, naked of apples only as high as the deer can reach on their hind legs. I took the frozen apples from the tops of these small trees back to the cabin, where I boiled, simmered, dried, crushed, and otherwise experimented with them.

Frozen streams also held many secrets. The deer knew where the water dropped over a rock at fantastic speed, deep beneath the snowdrifts. They would break through there for an icy drink, and then steam would rise from these holes because of the difference between water and air temperatures. Todd and I dipped a long, dry stick into the hole to make an instant popsicle. But we discovered that the ice would freeze solid to our tongues—a lesson hard learned.

Deer apples

Waterfall

In the spring, lamb's quarters was a favorite bedding for the deer. I also learned that jewelweed is a poison ivy antidote and as luscious a vegetable as lamb's quarters. Near the camp was a hundred-square-yard patch of jewelweed in which the deer fawned. What a place to give birth! Drops of crystal-clear water hung from the emerald-green stalks of the jewelweed as the deer gorged themselves on this plant. Level weed patches grew uniformly, so I could often tell where the deer have been by the scalloped designs of the weed tops.

The deer were sometimes invisible in this patch. As I crossed the area one day, a buck rose beside me, covered with cascading jewelweed stalks. We stared at each other, not more than three feet apart. For a long moment his antlers had looped over me like branches on a tree. I smelled his breath as he turned and walked away slowly. Through my eye contact, I apologized for my intrusion and swore I would call out loud before entering the area in the future.

Here's Mud in Your Eye

The homesteader in the Adirondacks needs to utilize both modern and ancient methods of survival when it comes to the insect world. For example, insects are particularly fierce during the short summer in the far north. Early in the day, black flies were biting even the corners of my eyes by 9:00 A.M. By noon, my upper arms were filled with itching lumps, the back of my neck was bleeding from my scratching, and I had swallowed at least four of the beasts.

I had literally used bottles of insect repellent, but even the heavy-duty spray from the Grand Union in Indian Lake was not working. The black fly problem was so great that all of us were sick from the toxins in our body from both the flies and the chemicals.

One time I came home to a prostrate husband, bloated and swollen from at least 400 black fly bites and sporting a temperature of over 103 degrees. A trip to the town doctor was in order. Since that time, we discovered the use of smudge pots. In fact, smudges soon became a skill we excelled in! Smoke wafted from our campfires, pails, upside down wastepaper baskets—you name it. The black flies were kept at bay.

Building a Smudge Pot

In the neighborhood dump area we were able to remove lidded tin cans, the 20-gallon type. Manufacturers distribute products in them. In the city I would ask local garages for oil cans, and restaurants for old lard cans. If all else fails, use an old wastepaper basket and make a thin lid to fit.

The can must be washed thoroughly and pre-burned for safety. Build a fire using paper kindling. Cover and wait until this burns out. The "clean" smudge pot needs a door in the base. Cutting would be simple using modern tools, but in the "homestead times" we used a nail and a rock. We placed holes side by side, much like you might open a can without a can opener. The door to a 20-gallon container might be 6 by 8 inches or even smaller. Don't cut the fourth side of the rectangle, and bend out the door. Smudge cans last for a season or more according to how much use the "hinge" side gets.

Smudge pot. Metal container with lid, rolls of moss.

Be sure the lid fits if you use a makeshift wastepaper basket or thin pail. Control your airflow by leaving the lid off, opening your door a couple of inches, and experiment with keeping a fire going. When you have coals, place a moist material over them and stand back. Your smudge is smudging. After a while, from the size of smoke, you can tell when to add more wood, open door and lid, or add more smudge material. When you go away, simply put the lid on and shut the door.

Smudge Materials
- Grass clippings (be sure you don't use toxic plants like poison ivy, oleanders or chemical sprayed lawn clippings).
- Moss- any kind works well.
- Wet Sawdust.
- Other material, such as newspapers soaked with water, wrung out and tightly waded up. the tighter the wad, the longer the smudge.

The Last Hunt

The transition from meat eater to vegetarian was a difficult one. Nevertheless, it seemed the best way to follow through on an environmentarian lifestyle.

One day my son Todd accompanied me hunting, the first time for him. I know now that hunting was probably a mistake, but I hadn't suspected the eight-year-old did not know where his meat came from. I performed my usual stone-throwing act, and shot the first rabbit that ran out from behind the lumber pile. Unfortunately, the rabbit rolled over, righted himself, and to my horror continued to run straight for us. I took careful aim and shot it just a few yards from where we stood. The rabbit fell, twitched, and then was still. Todd screamed, horrified at me. "Why did you do that?"

I heard myself saying, "We eat them for dinner. That's why we grow them."

He stood defiant and drew himself up into a proud, straight position and said vehemently, "Well, I don't have to eat them again, and I won't—ever!" Todd walked off in a rage. He meant every word he said, and that was the last kill I made for food from the mammal world. As it says in Genesis 1:29, "Behold, I have given you every herb bearing seed, which is upon the face of all the earth, and every tree, in which is the fruit of a tree yielding seed; to you it shall be for meat."

My theory now is that if we could see the reality of taking life, we would all be vegetarians. Happily, many youngsters are being introduced to this thought by activists or through the animal rights movement. This may well change people's eating habits in the future.

Foraging on the Way Home from Work

Although we were still learning to survive on our own, we needed to continue earning money for a while. The eight-and-a-half-mile stretch of Cedar River Road became a pathway for me to a job at the Cedar River House: a famous hotel often visited by dignitaries. The hotel needed housekeepers, so to earn extra money I worked there one challenging summer. The hotel no longer exists, but the trip to and from work is a treasured memory.

Sometimes I asked for a ride, but usually I welcomed the long trek home in order to sample the food along the way. I would consume large amounts of red clover, crabapples, and dandelions on these trips. Indeed, endurance and energy were sustained by these wild foods.

The deer became accustomed to my ways, and as I passed by them, they made no more than a faster twitch of their tails to acknowledge my presence.

I once saw a large bear along the road eating crabapples. He raked the apples from the branches of the tree with ease, then dropped down on all fours and pulled them up into his mouth. I watched him from a distance of a hundred yards, feeling very exposed. I waited more than twenty minutes for him to get his fill before I walked on. There was no need to disturb the feeding animal when I could avoid doing so, even though my family was waiting for their dinner.

Fire Making

I became an expert fire maker through daily repetition. We gathered kindling from around the area and found twigs for tinder underneath the branches of a balsam or pine tree. Birch bark became a favorite fire starter. Even through torrential rains and high winds, the birch bark always burned steadily.

There are many types of outdoor fires. My favorite fire pit was dug where there were no roots or rocks. It was located near our old homestead, with a rain barrel within a few feet in case of emergency. I used a shovel to sprinkle dirt over the fire when I had to leave it quickly.

The shanty, fire pit & wood pile.

After a few months, the dirt had been dug out and reused so many times that the ashes and dirt were a fine powder. Afraid of breathing the lye of hardwood ashes and dirt, I began to use a large piece of sheet metal, throwing it over the fire quickly. It wasn't long before I discovered this metal was a stove like the most modern type. I began to make roaring fires with the sheet metal on top, and I would simmer, boil, and fry all on one safe surface. In pouring rain or blizzards this method worked exceptionally well.

I found that a piece of sheet metal also served as a cover for a banked hardwood fire, and the coals would last until morning. If we went on a long foraging trip, the sheet metal cover helped prevent our coming back to a cold fire pit at dinnertime.

A fire pit holds a dug-out below-ground fire. A fire ring is an above ground safety rock ring around a fire area.

After a deep snowfall, we dug out the fire pit to prepare our meal and cooked in the middle of a snowstorm. During spring and early summer, we would cover ourselves with mud while cooking to protect ourselves from insects.

Scanning the coals using my hands, I could determine which areas of the fire would be used for cooking each dish. To control the boiling level I added twigs or hardwood to the fire. Since tending the fire and stirring the food are a constant chore when using an open fire, the children were often called into action. On Thanksgiving they spent eight hours turning a well-stuffed turkey over our fire pit.

My favorite type of fire ring was open on one end. I pushed the long logs in as I needed them; putting a new log on the fire means fresh fire or flames in the surrounding area. I made hotter areas for simmering or boiling by adding new twigs or fresh logs. The hot rocks on

Fire ring and cooking.
Open-end fire, with logs that are rolled over to control heat level.

the sides were used for toast or baking cakes. I greased the rock for best results as I would a pan.

To cook on our campfire, I used a spit for long-term roasting, coals for baking, and hot rocks for making toast. I baked main meals in coals (casseroles). Hot rocks and back burner held breads, which were unleavened, pita-like cakes throughout the wilderness years. Toast in particular tasted fantastic when done on the open fire. I would turn the bread when I saw the tiny coils of smoke appear at the edges, and I learned to keep the bread to the front edge of the grill to avoid burning the hair on my arms.

In the dead of winter I especially enjoyed cooking over the open fire. There were times when my back was thick with snow, but I was oblivious to discomfort. I would squat in front of the fire and relish preparing my family's meal. The personal challenges of foraging for and cooking our wild foods over our campfire gave us all a special appreciation of our daily meals.

Swamping

Our camp was bounded on the east by a vast cattail swamp and so the food became a mainstay.

"Swamping" brings out fantastic feelings. The marsh odors and insect life opens my senses and I feel animal-like. I made the mistake of wearing sneakers the first time I went out. "A little mud won't hurt me," I thought. Unfortunately, swamps and marshes have a tendency to suck the sneakers from your feet. As you step up, your shoe stays behind. And to retrieve it from the muck is no small feat. Marsh mud is thick, gooey, and smelly at the least.

Detractions aside, a number of edible plants can be found in marshes. All the plants described in this guide are fresh water plants. Mint and arrowhead are prolific, while meadowsweet and black birch also grow among the cattails.

When you go swamping, wipe a thin layer of mud on your face (temples to cheeks).

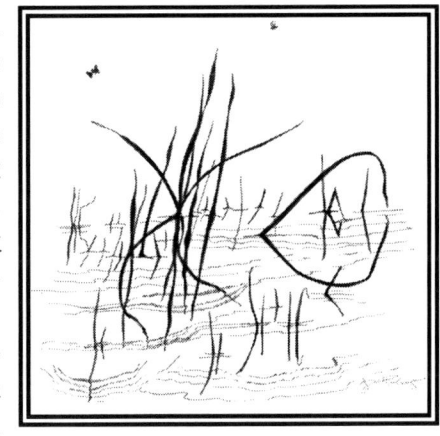

Swamp Grasses.

In fact, smear mud on as much skin surface as you can. This gives you a neutral marsh odor and bugs won't pay as much attention to you.

I would lay boards down on the marsh one in front of the other as a stabilizer. When I was ready to move on, I would pick up the first board and lay it down again in front of the second. You get to keep your shoes this way.

Using the stream to wash cattails

Water Hauling

Water is a basic need for all of us, but when you live in the city, you take it for granted. Our transition from city water to wilderness water was a monumental adjustment.

Ken did most of the water hauling, but there were many days when I shared the burden, carrying the five-gallon container on my back. The feeling was one of extreme satisfaction—filling, hefting, and toting the container to camp to supply our water needs for several days. I would use the drinking water one day and

The Wash Rock

A wash ring

then use the leftover water for cooking, washing, and preparing food. We also placed rain barrels in strategic places and washed ourselves with biodegradable soap when saponin (lather-producing) plants weren't available. At the river, everyone had his or her favorite spot to bathe in privacy. It was easiest to wash our clothes there, too.

Icicles

Under our uninsulated roof were the largest icicles I have ever seen. We used them for water. If we dropped one indoors, we squealed with delight as we chased the broken sections all around the linoleum floor of the camp kitchen. The average icicle was about four to six feet long and sometimes 12" or more in diameter. I could not leave them alone.

My son Todd helped me with this chore. Ice and snow melt at a great rate of speed in a bucket that is set into the top hole of the wood stove. A giant icicle might make four pails of water when melted. It was a true gift to the person washing clothes, dishes, wild foods or themselves.

Icicles

Reverance for the Wilds

The following poem was given to me by my mother, Ruth. It expresses my feelings for my time spent in the Adirondack Mountains. Its author is unknown, but the poem is from the early nineteenth century, reprinted from *Nirvana Lodge Log Book*, Indian Lake, New York.

A Camper's Prayer

God of the Hills, grant me Thy strength
to go back to the cities without faltering.
Strength to do my daily tasks
without tiring and with enthusiasm.
Strength to help my neighbor,
who has no hills to remember.
God of the Lake, grant me thy peace,
and Thy restfulness.
Peace to bring into the world of hurry and confusion.
Restfulness to carry to the tired ones,
whom I shall meet every day.
Content to do small things from littleness,
Self control for the unexpected emergency
and patience for the wearisome tasks.
With deep depths within my soul
to bear with me through crowded plans.
The brush of the night time where the pine trees
are dark against the sky.
The humbleness of the hills
who in their rightness know it now,
the laughter of the sunny days
to brighten the cheerless spots in a long winter.
Fill me with the breadth and depth and
height of "Thy Wilderness."
Thou has taught me by every thought
and word and deed.

The Simple Life

The following was written by Mrs. Barbara Jennings of Long Lake, New York, and is included here with permission.

Living amid the splendor and protection of the mountains, the people lead comparatively quiet lives. From year to year, very few changes take place, and everyone becomes accustomed to a routine. Consequently, the members of these small communities are independent and fixed in their ways. They keep well informed through television and newspapers, but because of the distance to be traveled, they don't often venture away from the local scene. The tourists are accepted as a necessary intrusion and the tourist season brings enough turmoil to last the entire year. Every man prefers to live according to his own pace, no one setting a very fast one. Those who maintain a carefree existence somehow seem to manage nearly as well as those who do work steadily and live moderately. Many prefer to have less, rather than submit to the shackles of a time clock.

On the occasions when we leave our towns, we are startled into a realization of how naive and backward we must seem. We may have read that skirts are shorter this season, but complete shock is in store when we arrive in the big city and discover how really short or long they are.

A friend and I felt quite at ease and well dressed when we arrived in Utica for a shopping trip, but a shoe salesman soon squelched me as he removed one of my best shoes and remarked, "Say, these are real old-timers, aren't they?" I spent the rest of the day trying to get lost in the crowds.

Even the ever-present desire of women to get out and explore the bargains and varieties offered in the large department stores soon diminish once they are there. They become entangled in the confusion of clutching hands at the sales counters and in the hopeless attempts to attract the attention of tired and bored sales ladies. Even the latest methods of self-service result only in standing in long lines waiting to be checked out. More often than not they return home weary and empty handed, resolving to remain a slave to the mail order catalogs and corner grocer rather than submit to such indignities.

Periodic trips to the bustling city are sufficient for anyone claiming to be a mountaineer by birth or by choice. They are ready to return to their towns where the roads may wind endlessly, but stoplights and traffic cops are nonexistent. No one waits in a booth ready to extract a toll. Even in this highly complicated and sophisticated age, the values of a simple life cannot be denied.

Left: Violets Right: Cattails

Foreground: lambs quarters

An Introduction to Edible Wild Plants

My life has been spent in two geographically opposite environments, giving rise to an environmentarian diet mixing woods & wilderness with Arizona desert. This experience is reflected here in the wild plants I choose to write about and the uses I've made of nature's bounty.

In many ways, the history of the United States documents people's efforts to survive. This is especially true of Native Americans, who spent long years establishing roots in this soil. In the Eastern wilderness, the Iroquois were bark eaters. Intense cold gave rise to the habit of eating fern roots, balsam needles, and pine bark, and of chewing black and white birch pith. Native Americans invented ways to dig roots from under the snow, such as for ferns and wintergreen.

In the West, specifically in what today is southern and central Arizona, the Hohokam were the nation's prehistoric people. From 300 BC to 1450 AD, they were hunters and gatherers. Wild foods were their diet, including cholla, lamb's quarters, amaranth, purslane, and saguaro.

Cross section of Arizona Desert by Willie Whitefeather

From these people came the desert Pimas, Navajos, and Hopis of the Southwest. Even today wild foods are a large part of the Papago Indian's diet, which includes milkweed, bulrushes, cattails, phragmities, amaranth, sumac, asters, thistle, sunflower, wild lettuce, chickweed, mustard, shepherd's purse, mint, plantain, sheep sorrel, purslane, willow, and rose (Hodgson 1982).

Just as Native Americans followed the patterns set by their environment, I found myself doing the things that nature made most obvious for me. I also studied the ways of people who had come before me.

Some specific foraging heritages I learned about are:

- Abanakee (People of the Bark)-eaters of birch, balsam, and willow
- Algonquin (People of the Coast)-eaters of amaranth and lamb's quarters
- Blackfoot, Chippewa (People of Canada)-rice eaters
- Cayuga (People at the Landing)-vegetable eaters
- Oneida (People of the Standing Rock)-vegetable eaters
- Objibwa (People above the Climate Line)-rice eaters
- Iroquois, Eastern Mohawk (People of the Flint & Keepers of the Eastern Door)-bark eaters
- West Seneca (People of the Great Hill & Keepers of the Central Fire)-bark and vegetable eaters

---

The point is, no matter where you live, you are going to draw upon the heritage of those who went before you. We can learn an immense amount from Native Americans. Climate, terrain, existing plant life, and economical use were all considerations of native peoples. The same plant might be used quite differently in different regions. To make use of the wild foods in your area, study your local Native American heritage. You'll find the information in local museums and libraries. Also, native teachers are often available, reached via earth gatherings, through organizations, at powwows, and such.

A Basic Food Source

Although some wild foods are best utilized in one particular way, such as steeped for tea, others have multiple uses and applications. In fact, it sometimes seems as if a single plant can meet just about all our needs. Such is the case with clover. We can eat it fresh or raw, or we can cook it, dry it, freeze it, or grind it into flour and bake it.

Clover: raw, cooked, dried, frozen or ground into flower.

The same can be said for amaranth, aster, chickweed, dandelion, dock (with some restrictions), lamb's quarters, mallow, nettles, plantain, purslane, shepherd's purse (with some restrictions), sow thistle, tumbleweed, and wild lettuce.

Gathering

The chart on the following pages delineates the uses of the wild foods discussed in this book.

Properties Of Wild Edible Foods

Plant	Annual	Biennial	Perennial	Native to N.A.	Naturalized	Harvest Time	Leaves/needles	Stems	Buds	Flowers/Catkins	Fruit	Seeds	Roots	Freezes	Dries	Vegetable/Fruit	Herb	Tea	Flour	Field Guide Page
Aloe Vera	○			○	○	Spring/Summer	○	○	○	○				○	○	○	○	○		52
Amaranth			○	○		Summer	○					○		○	○	○				54
Arrowhead			○	○		Fall							○	○	○	○				55
Aster				○		Fall	○	○	○	○		○			○	○				57
Balsam Fir			○	○		Year-round	○		○	○		○			○	○		○		58
Birch			○	○		Year-round		○	○	○	○	○		○	○	○		○		60
Blackberry			○	○		Year-round	○	○	○		○			○	○	○		○		62
Blueberry				○		Summer					○			○	○	○				64
Bulrush					○	Summer/Fall	○	○		○		○	○	○	○	○			○	65
Burdock		○				Year-round	○	○					○	○	○	○	○	○		67
Cattail			○	○		Year-round	○	○	○	○		○	○	○	○	○			○	69
Chamomile	○				○	Spring/Summer	○	○	○	○		○		○	○		○	○		72
Chickweed	○				○	Year-round	○	○	○	○		○		○	○	○	○	○		73
Chicory			○		○	Spring	○	○	○	○		○	○	○	○	○	○	○		75
Cholla			○	○		May		○	○	○	○			○	○	○				76
Clover			○		○	Year-round	○	○	○	○		○		○	○	○	○	○	○	78
Crabgrass			○		○	Year-round	○	○				○		○	○				○	80
Daisy			○		○	Year-round	○	○	○	○				○	○	○				81
Dandelion			○		○	Year-round	○	○	○	○			○	○	○	○	○	○		83
Dock			○	○		Spring/Fall	○	○				○		○	○	○			○	85
Evening Primrose		○		○		Summer/Fall	○	○		○		○	○	○	○	○				86
Filarie		○			○	Winter/Spring	○	○						○	○	○				87
Fireweed			○	○		Spring/Summer	○	○	○	○				○	○	○	○	○		89
Goldenrod			○	○		Spring/Sum/Fall	○	○	○	○		○		○	○		○	○	○	90
Grape			○	○		Fall	○				○	○		○	○	○				92

20

Plant				Season										Page
Lamb's Quarters	●		●	Summer/Fall	●	●		●	●	●			●	93
Malva		●	●	Winter/Spring	●	●		●	●	●		●	●	95
Maple		●	●	Summer	●	●		●	●	●		●	●	97
Meadowsweet		●		Summer	●				●	●				98
Milk Thistle	●	●		Spring/Summer	●	●	●		●	●		●	●	99
Milkweed		●		Spring/Sum/Fall	●	●			●	●				101
Mint	●			Spring/Fall	●		●		●	●		●		103
Mullein			●	Spring/Fall	●	●	●		●	●		●		105
Mustard	●		●	Spring	●		●		●	●				107
Nettles		●	●	Spring/Fall	●	●	●		●	●	●		●	108
Phragmities				Summer			●		●	●				109
Pine		●		Year-round	●	●	●	●	●	●		●		110
Plantain		●		Year-round	●	●	●		●	●		●		111
Prickly Pear		●		Summer	●	●		●	●	●				113
Purslane	●		●	Summer	●	●	●		●	●			●	115
Queen Anne's Lace	●			Summer	●	●	●	●	●	●		●		116
Raspberry		●	●	Summer	●	●	●		●	●		●		117
Rose		●	●	Spring/Fall	●	●	●		●	●		●		118
Saguaro		●		Summer	●	●	●		●	●				120
Sheep (Garden) Sorrel	●			Spring/Summer	●	●	●		●	●		●		121
Shepherd's Purse	●		●	Summer	●		●		●	●				122
Sow Thistle	●		●	Winter/Spring	●	●	●		●	●			●	124
Strawberry		●		Spring	●		●		●	●		●	●	125
Sumac		●		Summer	●				●	●		●		126
Sunflower	●			Spring/Summer		●			●	●		●		127
Thistle		●		Spring/Fall	●		●		●	●	●			128
Thyme		●		Summer/Fall	●		●		●	●		●		129
Tumbleweed	●		●	Spring/Summer	●	●	●		●	●			●	131
Violet	●			Spring/Summer	●	●	●	●	●	●	●	●		132
Wild Lettuce		●		Spring/Summer	●	●			●	●		●	●	133
Willow		●		Year-round	●	●			●	●			●	134
Wintergreen		●		Spring/Summer	●				●	●		●		135
Wood Sorrel		●	●	Spring/Fall	●	●			●	●	●			137
Yarrow		●	●	Spring/Fall	●				●	●		●	●	138

Picking Wild Foods

State and local regulations on picking wild plants vary, but mostly there are restrictions against taking plants or plant parts on public land. You should always check with park officials before foraging anywhere but on your own land.

In gathering plants, pick only on lands off non-posted access roads or from fields with owner's permission. Move from place to place, picking sparingly and only when there is several of any one plant. In addition: due to nature's changes and people's advances, many plants are now endangered. Even though this guide's plants are not endangered now, they may be in the future. Check your Local State library or Department of Agriculture for updates. Do not pick any endangered plants on private property.

FORAGING RULES

1. **DO NOT** collect plants closer than 200 feet from a road.
2. **NEVER** collect from areas sprayed with herbicides, pesticides, or other chemicals.
3. **ALWAYS** be familiar with all dangerous plants in the area.
4. **POSITIVELY IDENTIFY** all plants you intend to use for food or medicine. Check against three good field references with excellent illustrations.

If you are trying a plant for the first time, after checking your field guides, do the following:

1. Snip a piece of the plant and roll between your fingers and sniff. Discard if objectionable. If you like the smell, then rub the tiny piece on your GUMS, above your teeth.
2. Wait twenty minutes.
3. CHECK For burning, nausea, stinging, itching (all allergy results). Poisonous plants USUALLY produce one or more of these symptoms.
4. If no untoward reaction results, take another tiny bit of the plant and make a weak tea. (Place piece in teacup, pour boiling water over, cover, and steep for 10 minutes. Ingest a <u>small</u> amount.)

5. Wait another twenty minutes! Check for signs of irritation. If none, then reheat the tea and sip slowly.
6. Keep all samples away from children and pets. Keep edibles separate from samples to be identified—poisons will give their bad qualities to food through contamination. Bag foods, poisons, and samples separately.
7. Be aware that heating or boiling does not always destroy toxicity.

In general, it's a good idea to store all seeds and bulbs away from children and pets. Tape down the lids or use a locked closet for any questionable foods so children cannot reach them. Likewise, teach children to keep all plants away from their mouths. Do not let children chew on or suck nectar from unknown plants.

Avoid smoke from burning plants. Smoke may irritate the eyes or cause allergic reactions. Also, be aware of your neighbor's habits with regard to chemicals, pesticides, and herbicides. Ask questions! Call and report any chemical spills, contaminated areas, or other situations that may negatively affect your gathering safety.

Growing up with Wild Foods

"As a child, I always mashed and ate wild plants. This worried my Daddy. He began to teach us what not to eat and what good plants to enjoy. He showed us how to pick flowers from the top of ocotillo plants while on horseback. We sucked the honey from the bottom of each flower, much as Easterners used honeysuckle. The ocotillo flowers are just delicious. Our parents ran a cow ranch in Mohave County for fifty years, and we picked our ocotillo flowers often from horseback. A lovely life!"—Danielle Stephens, Kingman, Arizona.

Gathering

Decide which plant you will harvest on a particular day and take along a basket, a bag, or a sheet. Burlap is excellent if you are going to dry the plants, but biodegradable plastic bags generally are better for gathering because burlap gets heavy when wet.

Do any aquatic plant gathering in the morning, as the cool air and water will keep your collection fresher. Plastic bags are waterproof and lightweight, but remember to use biodegradable bags, and reuse them as often as possible. (Do not leave wild food in the sun, especially inside a plastic bag, because the food will wilt quickly; I have spoiled hundreds of milkweed buds this way!)

A flat, round tray also works well for gathering in tall grass, especially when gathering fireweed. It can be great fun wading through a field of these magenta flowers, snapping off the tips and tossing them onto a tray held above your head.

For most gathering, I use a clean sheet and a two-foot section of string. A small amount of wild food will fit in your basket or bag, while a winter's supply will fill a clean sheet. When the sheet is folded up and tied with a string, it can be hung up easily in a warm, airy spot to dry. The sheet method of collection also allows the bugs to crawl out of the food and provides an easy view of how much you have gathered.

When gathering wild foods, remember that certain plants snap off easily while others are better cut cleanly with a knife or scissors. If the plant gives you any resistance to a snap, give way to a tool immediately. A clean cut will help the plant recover quicker and you'll have greater gathering speed and accuracy.

Cutting a batch of leaves on the stem is simple when you hold the stem with one hand and with the other cut the stem with a sharp knife. When I gather with a penknife or scissors, I slide my fingers into the middle of the clump and go for inside growth. The plant then folds up around what I took and rejuvenates nicely. Goldenrod, asters, yarrow, and Queen Anne's lace work especially well this way.

If you're using the sheet method, you may want to strip the leaves off the stems while gathering for ease in drying. At this time you can also separate your favorite flowers for tea. Don't forget to save the stems, though, as they can be valuable, especially mint, spearmint, yarrow, and thyme.

When you've gathered enough, pull the corners of the sheet together, and use the string to tie it closed. I often hang the bundle from a tree or other obvious spot and retrieve it later. This method also helps mark the trail. (I use a cactus in the desert since trees are sparse.)

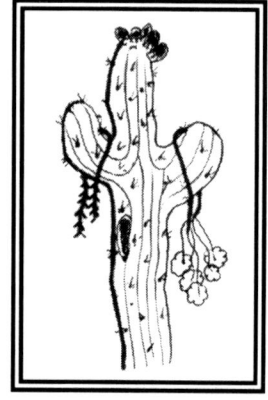

A Food Tree   **Saguaro as a food tree**

Tie enough stems together to make a bunch, but be sure the leaves aren't packed too tight, lest you prevent air penetration. As a rule, when you swish the bundle in the air you should "feel" and hear the air moving through it.

Waists and belt loops also work well for carrying bundles of plants. You can also use burdock leaves or mullein to roll up several smaller items. Roll up nuts inside large leaves and put them in your pockets as you go. Fruits, twigs, nuts and flowers all wrap well in large leaves.

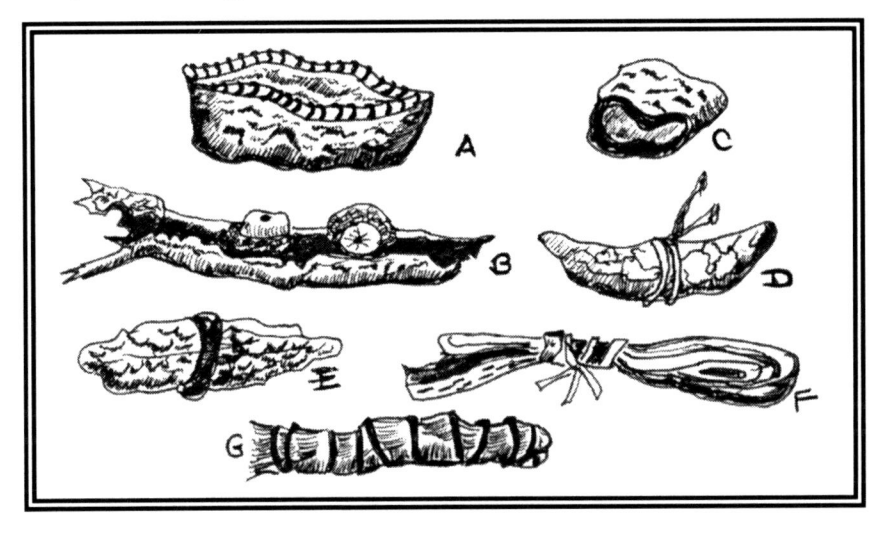

Leaf Bundles: A) White birch bark waterproof container, B) A bark tray, C) Hollow bark, D) Burdock leaf wrapping for small foods, E) Mullein leaf wrapping for small objects, F) Kit of white birch strings, and G) Bundle of foods, grasses or twigs.

In summary, here is a list of tools I generally use for gathering:
- A sheet
- Scissors
- 2-foot length of string
- Biodegradable plastic bags
- Penknife and Bowie knife
- Gloves
- Snake stick with fluorescent handle (in the West)
- Hand axe (in the East)
- Burlap bag

Washing the Foods

Certain plants attract aphids, ants, and "no-see-ums," as well as collect the usual dirt and dust. In the Adirondacks a nearby stream is always available, so washing large quantities of plants is always easy, no matter how dirty they are. But in Arizona, apartment living demands the use of a bathtub—it's the easiest way to wash a large quantity of food. Of course, it's always best to use a hose first, especially for roots, eliminating most of the soil before taking them into the house.

When cleaning small plants or bunches of leaves that aren't too dirty, it's a good idea to use a drainable container such as a colander. Some containers can be made to drain by poking holes in the bottom. Some plants may need more soaking, however, and may even need to soak overnight. In this case, soak in 1/4 cup vinegar to a tubful of water. In any case, it's best not to put too many types of plants in the same tub, since you may have difficulty separating them later.

Finally, here's a quick tip that will make it easier to clean your tub and prevent a clogged drain: Before filling the tub, place a piece of cheesecloth over a rubber stopper and use it as the plug. When you're ready to drain the tub, pull the stopper, but hold the cheesecloth in place. When the water is drained, just shake out the cheesecloth and save it for the next time!

Edible Wild Flowers

The following are common wild flowers I utilize often.
Birch catkins—unopened catkins eaten raw or brewed as tea. Dried or frozen for storage. Also cooked.
Blue Asters—All wild asters and flower buds may be eaten raw, cooked or dried and ground for flour.
Cattails—Flowers and flower buds are eaten raw, cooked, or dried and ground for flour. Used in tea.
Chamomile—White flowers and flower buds may be eaten raw, cooked, steeped for tea, or dried for storage.
Clover, Red or White—Flowers and flower buds are eaten raw, cooked, dried for tea, & frozen in ice cubes.
Dandelions—Whole flower and flower buds may be eaten raw in salads, stir-fried in garlic, frozen for use later. Used in tea.

Fireweed—Whole flower and flower buds are eaten raw, cooked, or dried for storage.

Goldenrod—Whole flower and flower buds are eaten raw as a tonic, and dried for tea. Frozen for storage; used in tea.

Maple—Flower buds are eaten raw, cooked, or dried for tea and storage. Can also be frozen.

Meadowsweet—Flowers and buds are eaten raw, cooked, or dried for a sugar substitute or tea.

Milkweed—Eastern flower is cooked as a vegetable, or frozen or dried for storage. Western varieties are caustic from desert sand.

Mint—Flowers and flower buds are eaten raw or cooked, or dried for storage; frozen in ice cubes; used in tea.

Mustard—Flowers and flower buds vary in intensity; may be cooked or eaten raw, and frozen.

Pine—Pine flowers or young cones are called catkins and are eaten raw, cooked, and dried for tea; also frozen.

Prickly Pear—Flowers are eaten raw in salads, and cooked; frozen for storage.

Queen Anne's Lace—Flowers and buds are eaten raw, cooked, frozen, and dried for storage.

Roses—Flowers and flower buds are eaten raw in salads, cooked, also frozen in cubes for storage.

Thyme—Flowers and buds are eaten raw, cooked, or dried for tea; also frozen.

Violets—Flowers and flower buds are eaten raw in salads, cooked, and dried or frozen for storage and tea.

Wintercress—Flowers and buds are eaten raw, cooked, and dried.

For information sections on drying and freezing flowers, see pages 33, 37, 39, 42 and 43 respectively.

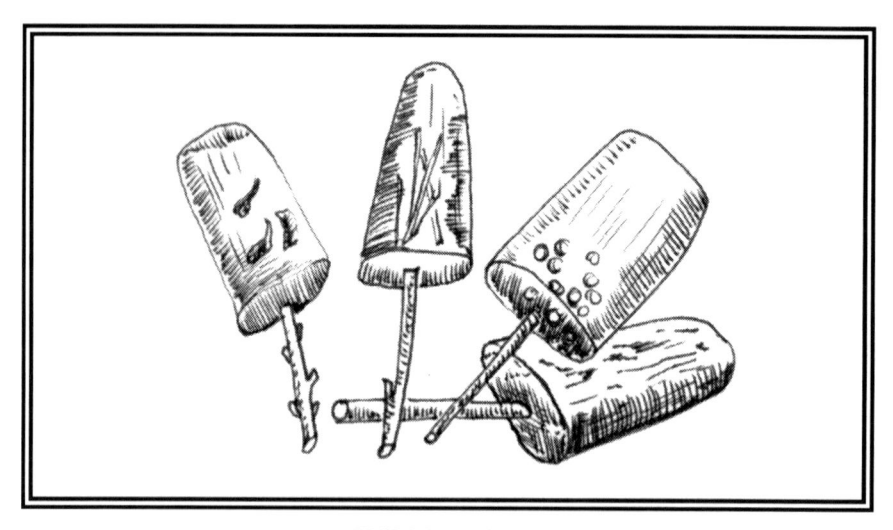

Wild Popsicles

Edible Roots

During my homesteading days I became interested in the many uses of plant roots. A particular experience gave me a lasting respect for the roots of wild plants.

In mid-winter, the ground was frozen and a snow-encrusted world was our environment. My son Todd had developed a bad cough. Soon Todd's condition worsened and he was coughing steadily. He needed more than the usual remedy of pine tea and honey. I knew about the antihistamine properties of mullein root. I could see the tall brown spikes of mullein poking above the snow like sentries standing beyond our camp. But the temperature was 20 degrees below zero and the ground was frozen solid.

The woodstove was filled with hardwood coals. I collected a pail of coals, put on my snowshoes, and went out to make a path to the closest mullein plant I could find. I scraped away as much snow as I could from around the plant and poured the hot coals on the icy crust at its base. (A very heavy pail is necessary for this, as coals will burn through a thin pail very quickly.) Sizzle, sizzle. The coals sunk quickly into the crusty ice. I returned to the house and fetched more coals. Pausing a few minutes between pailfuls, I scraped away each layer of slush until I reached mud. I placed a circle of dry wood around the base of the plant and made a large campfire inside the circle. With the wood concentrating the heat at the plant's base, the fire became hot in no time. I used birch bark to keep it very hot, a large sheet or two at a time. We were lucky to live nex. sawmill that cut white birch logs, and sheets of bark were ple. ful. Never cut birch strips off a live tree.

After a long, cold hour, I scraped the remaining coals out from the circle and easily dug out a foot-long mullein root. It was partly decomposed, but there was more than enough root to brew a concentrated tea for Todd. He drank 2 tablespoons every hour, and his cough was calmed. Mullein root tea is an effective expectorant, helping Todd discharge mucus from his throat and bronchial tubes. This wintertime experience gave me untold security, knowing that mullein roots were out there if we needed them for medicine or food. This "underground farm" took on new meaning: it was available to us even through several feet of snow and in 20 below temperature.

The following edible roots are ones I use:
Amaranth—Used as a vegetable; dried & ground for flour.
Arrowhead—Tuber used as potatoes.
Balsam fir—Used for medicinal purposes (sedative), & tonic.
Burdock—Used as a vegetable, dried and ground for flour.
Cattail—Dried and ground for flour.
Chicory—Used as a coffee substitute.
Dandelion—Used as a vegetable and tonic.
Evening Primrose—Used as a vegetable flour.
Fireweed—Used as a medicine and a food.
Goldenrod—Used for a blood tonic tea.
Malva neglecta—Used as vegetable; dried & ground for flour.
Pine—Used for tonic and stimulant as well as rope making.
Plantain—Used for vegetable and as an antiseptic
Purslane—Used as a vegetable; dried and ground for flour.
Queen Anne's Lace—Used as a vegetable and dried food.
Strawberry—Used as a vegetable, a medicine and as an appetite stimulant.
Wild lettuce—Used as a vegetable.

Plant roots are greatly under utilized, and their magnificent properties barely understood. For example, many are aromatics.

Roses, for instance, have an aromatic root that can be reused many times, until the scent and color are gone. Scrub off

ick spots before use. Mullein root can be reused
ea, as can mint roots.
ly gathered roots from the plant and scrub with
then use fresh, or dry and store.
e fresh, peel the root as you would a carrot or slice crosswise or lengthwise. To dry the roots, place bag and label carefully with the name of the plant and year you collected. Hang to dry. Reconstitute in boiling water or place in a grinder and grind to flour.

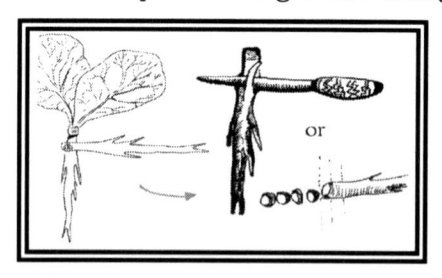

Harvesting roots: peel & slice either crosswise or lengthwise.

Storing roots: A) hang on wall B) dry & store in paper bag C) store dried slices in glass jars

Edible Parts of Trees and Shrubs

Many trees and shrubs have edible parts and can be a major source of food in the wilderness. The chemical content, vitamins and minerals are found in leaves and flowers, but in concentrated form in shrubs and tree bark. Inner bark, the layer underneath bark, is the "meat" of a tree. Twigs have both bark and inner bark. For example, pine or balsam fir bark strips, cut 1/3 inch to 1 inch wide, to fill a 2-quart glass container will yield a winter's supply of tea. One small shrub, such as a rose bush, can yield a winter's supply of tea.

In calculating your yield, figure one twig, 6 to 8 inches long, yields a moderately strong cup of tea. This tea could be used as a soup base. To prepare for cooking or tea, break the twig several times to release the flavor and steep or boil as intended. To store twigs and stems, place in dark paper bag. Or, place in glass jar and cap tightly. Wrap with opaque paper or foil to keep it dark. And, of course, label with the collection date.

The following diagram shows two different methods of peeling bark. Method A shows types of plants which will allow big pieces to be peeled. Method B makes smaller wispy pieces by first splitting along the length, then peeling small pieces around the girth from one side of the split to the other.

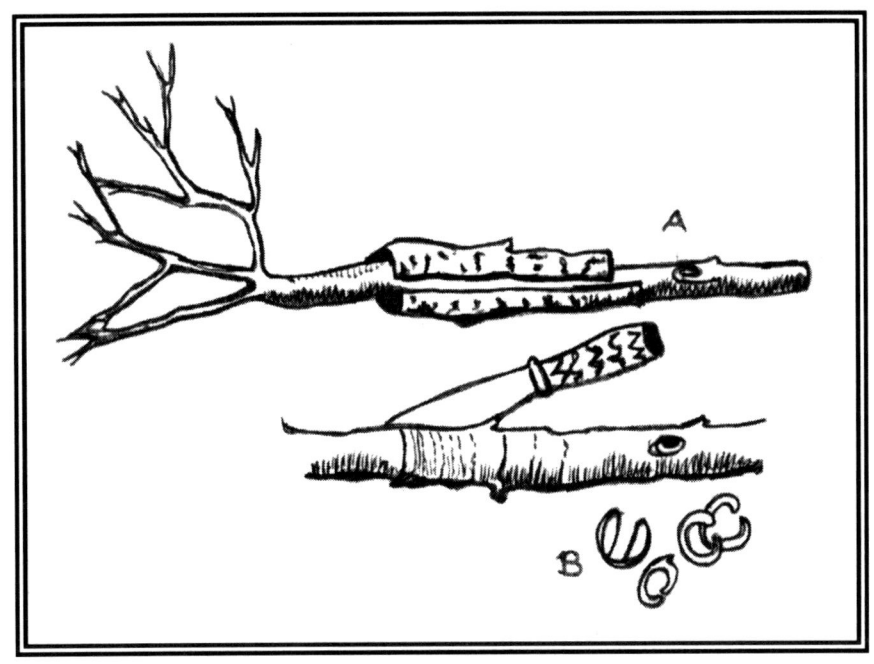

Peeling bark in strips.

The following are trees and shrubs and commonly harvested for their edible parts:

- Balsam fir—Parts of the entire tree are edible, especially the inner bark, sap, twigs, needles, and flower buds and seeds.
- Birch—Entire tree is high in mineral content and medicinal value; sap, buds, catkins and seeds are edible. Birch catkins, buds, twigs and bark, as well as inner bark, are all loaded with trace minerals, and beta-carotene.
- Blackberry—Inner bark, twigs, flowers, fruits and seeds are edible.
- Blueberry—The twigs, bark and roots are edible as well as fruits.
- Maple—Entire tree is edible, especially inner bark, sap, twigs, buds, young leaves, and seeds.
- Meadowsweet—Entire plant is high in minerals and medicinal value; inner bark, twigs, buds and flowers, leaves and seeds are not to be eaten in bulk.

- Pine—Entire tree is edible; inner bark, sap, twigs, needles, catkins, and in one species, Pinon pine nuts.
- Prickly pear—Buds, flowers, pads and fruits of plant are edible.
- Raspberry—Entire shrub is edible, including inner bark, sap, twigs, buds, flowers, fruits and seeds.
- Rose—Entire shrub is edible, especially inner bark, sap, twigs, buds and flowers, leaves and seeds (hips).
- Saguaro—Buds, flowers and fruits, and seeds are edible.
- Sumac—Flowers, fruits and seeds are edible.
- Tumbleweed—Twigs, buds, and leaves are edible.
- Willow—Entire tree has high mineral content or medicinal value and should not be eaten in bulk. Inner bark, sap, twigs, buds and catkins, leaves and seeds are eaten in moderation.

Teas from Wild Plants

Gathering and preparing teas from wild plants is simple and rewarding. For example, a large supply of thyme for tea can be harvested in just two minutes with scissors. Drying requires no more than a screen and takes but a few hours. Using the back shelf of my car, I can have it dried and ready for storage in a few hours. In a few days, the patch grows back and I can harvest it again. This is true for most tea plants.

Plants commonly gathered for tea include:

- balsam fir
- birch
- blackberry
- blueberry
- chamomile
- chicory
- clover
- dandelion
- fireweed
- goldenrod
- meadowsweet
- mint
- mullein
- raspberry
- strawberry
- thyme
- violet
- willow
- wintergreen.

(See Part II, the Field Guide, for information on individual plants.)

Picking Nettles

Do you want to try nettles but are afraid to pick them? Take your gloves, a pair of scissors, and a container and head for the nettle patch. Cut the leaves directly into the container, holding the plant with your gloves. Or, to take home the entire nettle plant, cut two individuals of a plant that is long and straight, such as goldenrod. Lay them on the ground parallel like railroad tracks. Cut the nettle, using your gloves and scissors, pile them across the top. Then pull the goldenrod stems up and fold over so you can carry your bundle home without getting pricked by stickers.

Storing Teas

You might enjoy fresh tea brewed from raw materials. Remember that raw material is a bit less effective and tastes weaker than dried.

Wild plants or leaves that have dried until crisp can easily be stripped clean for storage. Place a clean sheet on a counter and put bundles on top. Using gloved hands, pull the dried leaves, seeds, or flowers down gently, then scoop up and crumble by twisting the sheet and wringing it. This is what is known as a rough grind. In some cases you can obtain the same results by using your hands. Store rough ground teas in glass jars.

You may store twigs in "tea sizes", such as six inch sections of willow equals approximately 2 aspirin. Six inches of a rose stem equals 1 cup of tea. Four inches of goldenrod stem equals one cup of tea. If you use a mayonnaise or quart size jar many cups of tea are easily stored.

Window food dryer with removable racks.

Flour from Wild Grains

Bulrushes and phragmities grow prolifically in wet lands. They are a familiar sight to most of us, yet few people realize what a valuable food source they are. In both the East and West, I have gathered and ground bulrushes to make flour. The flour makes delicious muffins, cookies, and breads. What's more, there are common grasses growing in most people's lawns that can likewise be harvested for flour. My favorites are listed on page 35.

Gathering bulrush stalks is an easy task, so long as the water is clean. Know where the water comes from and look carefully at the stems as you cut them. It is a good idea to wear gloves until you are well used to spotting the stems of the rushes. Poison hemlock may be in the same marsh, so be careful! (See Part II, Field Guide, and Part V, Poisonous Look-Alikes.)

Bullrush

Cut one of the two stems of a bulrush plant at a time, sliding your gloved finger down the stem, cutting, and placing the rush up and under your armpit. As you move along, the batch of plants under your arm will grow. When you have enough, stop and tie them into a bundle. Hang the bundle in a tree or make a yoke for around your shoulders. Pick from the centers of the bulrush clumps, so the plant will grow back quickly.

Dry bulrush in an attic, car, or other warm and dry area. Let dry thoroughly. They may appear to be dry, but check. The stems break, crack or snap easily when dry. Place in glass containers until you can grind the grain to flour.

When you are ready to grind to flour, check bulrushes for tiny black dots. These may represent ergot, a poisonous fungus. If present, discard the plants.

Each bulrush stem yields several teaspoons of flour, and a screen full of stems and flowers yields approximately 1 cup of finely ground flour. The color of the flour depends on the bulrush: it will be green if bulrushes are picked early in the season, tan if picked later.

> *"He causeth the grass to grow for the cattle, and the herbs for the service to man; that he may bring forth food out of the earth."*
> Psalm 104:14
>
> *"Also, to every beast of the earth, to every bird of the air, and to everything that creeps on the earth, in which there is life, I have given every green herb for food, and it was so."*
> Genesis 1:30

Harvesting Grasses

The following are common lawn grasses that make excellent flour. They are also known as Graminae, since they are in that family of plants. Since these plants are not described in Part II, check identification in several other reliable field guides. The grains may also be cooked whole for granola, cereal, and other uses.

Barnyard grass— *Echinochloa crus-galli*, Graminae Family
Barley grass— *Hordeum pusillum nott*, Graminae Family
Broomsedge— *Andropogon virginicus*, Graminae Family
Crabgrass— *Digitaria sanguinalis*, Graminae Family
Foxtail grass— *Setaria italica*, Graminae Family
Goosegrass— *Eleusine indica*, Graminae Family
Jungle grass— *Echinochloa colonum*, Graminae Family
Quackgrass— *Agropyron repens*, Graminae Family
Rye grass— *Lolium temulentum*, Graminae Family
Wild oats— *Avena fatua, A. barbara*, Graminae Family
Wild rice— *Zizania palustris*, Graminae Family
Wheat grass— *Triticum aestiuum*, Graminae Family
Yellow nutsedge—*Cyperus esculentus*, Graminae Family

Some grains are better sheared with scissors, such as crabgrass. Others are easily collected using a pillow case. If the seeds come off the stem easily, just put the pillow case over the tops, bend the grasses over, and wring the seeds off. This method is especially good for marsh grasses. I've happily spent many an evening gathering a quart or two of seed heads while watching the dragonflies in the setting sun.

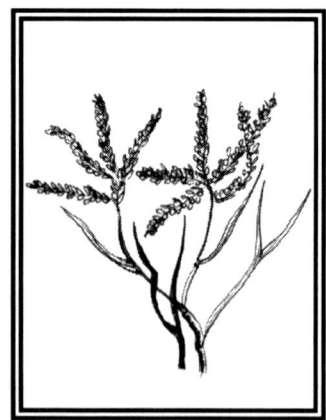

Barnyard Grass, Goose Grass & Wild Oats

After the grains are dried, they can be ground to flour. Use your own method; grinders, processors, and mortar and pestle are all successful.

I usually clip stems to six inches from the grain. Grinding a bit of the stem adds to the fiber content.

Grains contain protein, and protein is one of the building blocks of life. Unrefined grains are best because the natural elements have not been processed out. Much of wheat's protein and vitamins E and B^6 are lost in the refining process. As a result we have "enriched" bread, which tries to make up for the loss. Isn't it a better idea to keep those nutrients to begin with? This is why whole wheat flour and whole natural grains are the wise nutritional choice today. Also, fiber is lost in the processing of wheat flour. Yet the higher the fiber content of our food, the better for our digestive system. Unrefined, natural grains are truly the only answer to better health.

Blend one-half cup of assorted wild grains putting them into a wide-mouth thermos and then pour boiling water over them. Cap tightly and let sit overnight to create an instant hot cereal for the morning or the ingredients for a rough bread dough or wild "pancake." Add a little vegetable flour such as amaranth or clover for a high-energy meal!

Preserving Nature's Bounty

Drying for tea or grinding to flour, as the preceding sections describe, are major ways to preserve and enjoy wild food, but there are other possibilities as well. You can collect roots, branches, needles, and flowers in season and either dry them or freeze them for later use in a variety of ways.

In the Adirondacks, my pantry was full of wild foods preserved for winter. The assortment varied from year to year, and when I moved West, there were many substitutes. Tumbleweed took the place of fireweed. Malva neglecta was used instead of arrowhead for flour. The caption to the illustration on the next page tells what wild foods I commonly had in my pantry, East and West.

Typical Pantry Storage

Eastern cupboards were always filled with certain staples. I did not own a freezer in those years. All staples were dried.

Car drying racks.

Plants hanging in storage pantry.

The Pantry

<u>**Gallon jars of leaves included**</u>: Amaranth, asters, lamb's quarter, strawberry, clover, red and white clover, chicory, daisy leaves, dandelion, dock & malva neglecta. Western jars also included tumbleweed, mesquite beans, palo verde beans & chaparral.
<u>**Gallon jars of twigs for teas included:**</u> Rose, meadowsweet, pine, balsam, birch, maple & willow, and, in the West, ephedra (Mormon tea).
<u>**Separate jars of dried flowers:**</u> Aster, clover, chamomile, chicory, dandelion and rose, both in the East and West.

My closets always had a few choice branches stored whole: pine, balsam, willow, and maple. Closets in both East and West are filled with paper cartons of grasses for future use as flour and, of course, the Christmas tree and wreath from year to year!

Preparing Food for Storage

Once you've gathered your bounty, shake off any soil and wash plants, leaves or roots gently under running water. Wash all foods except cattail pollen. Scrub roots with a toothbrush.

Trim off any dead leaves or parts. Remove and separate leaves, seeds, flowers, and roots for separate drying; or leave plant whole (as you might dry lamb's quarters or amaranth for flour). Cut stems to storage-container lengths.

Drying Methods

There are a few basic "don'ts" about drying food that you should know before you begin. Don't dry leaves in direct sunlight; find a warm, airy place out of the sun. Never dry foods near a road or garage, or you'll taste exhaust fumes. Lastly, don't dry pine needles in an oven or microwave. They are very volatile and may explode.

The best indoor drying apparatus I know is a window screen, convenient for either an apartment or a home. You simply bring home your favorite wild foods, wash them well, trim, and shake off excess water. Lay the foods on the screens, keeping any branches apart somewhat. Slide the screen into bracket and start the next screen. (See page 33 for an illustration of screen assembly.)

When you feel that food is as dry as it can be, preheat the oven to 300 degrees and turn off. Heat the foods for five minutes, then put in hot clean jars, or put the gatherings in the jar, without the lid and heat jar, foods, and all for five minutes at 300 degrees. Often I simply turn the oven to 300 for a while, put the food in, and turn off the oven, leaving it to cool slowly. I take the jar out when the oven is partly cool, then cap tightly for storage.

There is another method. Although you may feel odd driving around town with a rack of "weeds" in the back of your car, it is by far the most efficient drying rack! In July and August, one day on the back window shelf of your car will thoroughly dry most leaves. Just think, while spending money driving around town, you are performing an energy-saving task and making a step toward long-term food storage, ultimately saving dollars. The car doesn't have to be in the hot sun all the time. Park in the shade if you wish.

Caution: Don't dry foods in a car when the outside temperature is over 80 degrees—toxic fumes may arise from plastic or vinyl seats.

The Drying Roof

There are more ways to dry wild plants. Here's a simple method I used.

Looking around for a flat place to put the screens I used to dry plants on, I noticed that the roof seemed to take up a lot of space—space that didn't do much other than keep the rain out. So I climbed a ladder and began to place the screens on the roof. Since there were poplars surrounding the cabin, the shade was perfect for drying apples and berries. We even nailed on boards as scaffolding to create a large accessible drying space. You may not want to go this far, but with a little ingenuity you'll find almost any place can be used to dry fruits, berries, and such.

We dried several types of wild apples. We did this every year using the larger varieties, choosing about half the normal size apple. We also dried crabapples on the roof. We gathered the crabapples by spreading a sheet under the tree and shaking the limbs, making fruit fall. We cored the apples and then sliced them crosswise about 1/3 inch thick. We threaded these slices on old broom handles, then hoisted them up to the roof. After a few days, the crabapples had shriveled and became leathery. After a 5-minute baking in the oven to dry them completely (oven preheated to 300 F. then turned off), we stored them in glass jars. In winter we used them in everything from cereal to cobblers, pies, and snacks.

> ***Linda says-*** You can also string the dried apple slices together with a needle and string. Not unlike the garlic and hot peppers in the Southwest, your wild foods can decorate your kitchen. I even string roots this way.

Label the jars with the date of drying before storing. The storage place is important. Closets work best. You need darkened spaces for dried foods, out of direct sunlight.

Store whole leaves and grind them into flour as you need it. For long-term storage, grind and place the flour in a shallow 1-inch pan and toast for five to ten minutes in a preheated 300 degree oven turned off. Store in glass for up to several weeks. Vegetable flour does not have a long life, so avoid storing for more than six months at a time.

Rinse dried branches to clean off any dust before you use them. Rose or raspberry vines, pine or balsam branches, and weeping willow can be stored for years, but will accumulate dust.

Containers for Dried Foods

Generally, I'll save just about any glass jar and use it to store dried foods. As long as they are in glass with a tight-fitting lid, wild foods can be stored for long periods of time. My friends bring me their glass jars: in fact, I have begun to consider myself a recycling center! Of course, brown glass is best because it lets in less light, but it is not readily available. You can find green jars with narrow screw tops, however, and these are ideal for storing seeds, nuts, and berries—things that pour well from a narrow top.

The best container for large amounts of stored leaves, twigs, branches, and such are the half-gallon glass containers that deviled eggs, pickles, and mayonnaise come in. Restaurants or convenience stores might save them for you.

Devices that seal plastic bags, removing the air, work almost as well for long-term storage. Roots store well in paper bags, as well as nuts, twigs, barks, and branches to a lesser degree. As long as food is as dry as possible, long-term storage is possible.

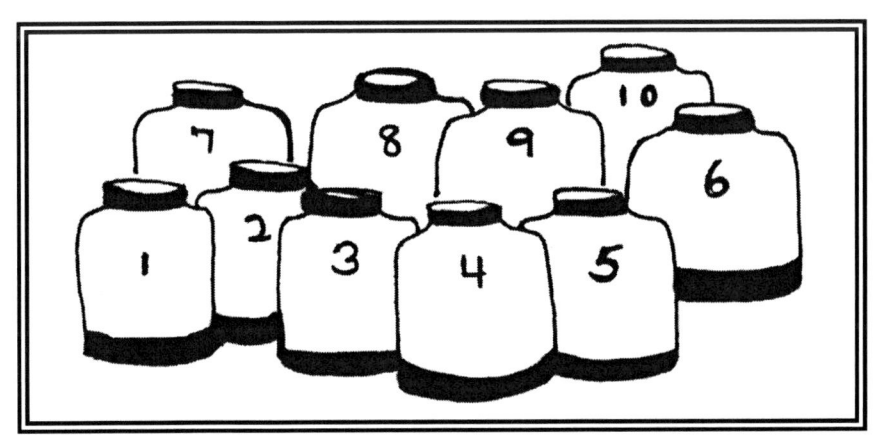

Storage Jars: #1 vegetables, #2 flowers, #3 twigs, #4 roots, #5 stems, #6 nuts, #7 corms, #8 buds, #10 flour

Storage Problems and Solutions

Problem: Pine needles that are stored in glass begin to display black spots or black mold.

This is known as ergot and develops on incompletely dried wild foods which are then sealed in airtight containers. Discard the contaminated foods as these molds are highly toxic. Similarly, grain that was incompletely dried also hosts this toxic black mold. Willow branches stored in a damp area outside develop black spots, also recognized as a poisonous mold.

Freezing Your Wild Foods

Freezer packages of vegetable matter have a life of six months. Roots, twigs and tubers last a bit longer as a rule. When I find freezer crystals throughout the freezer package, it's time for a massive green soup! Then you may re-freeze the soup.

The following are the wild foods successfully frozen. All must be steamed the indicated time before freezing.

Food	**Steam Time**	**Food**	**Steam Time**
Aloe Vera	raw	Plantain	1-2 minutes
Amaranth	1-2 minutes	Prickly pear	2-3 minutes
Arrowhead	1-2 minutes	Purslane	1-2 minutes
Aster	3 minutes	Malva	
Blackberry	2-3 minutes	neglecta leaf	1-2 minutes
Blueberries	1-2 minutes	Maple	2-3 minutes
Burdock	3 minutes	Milkweed	
Cattail	3 minutes	heads	3 minutes
Chamomile	1-2 minutes	leaves	2 minutes
Chickweed	1 minute	Mint	1-2 minutes
Chicory	3 minutes	Queen Anne's	1 minute
Clover	1-2 minutes	Raspberry	2-3 minutes
Daisy leaves	1-2 minutes	Rose	1 minute
Dandelion	3 minutes	Saguaro buds	3 minutes
Dock	3 minutes	Shep.'s purse	1-2 minutes
Evening		Strawberry	
primrose root	1 minute	leaves	2 minutes
Filarie	1-2 minutes	Thistles (all)	1-2 minutes
Fireweed	3 minutes	Thyme	1 minute
Goldenrod	1-2 minutes	Tumbleweed	1 minute
Grape	2 minutes	Violets	1 minute
Lamb's quarters	3 minutes	Wild lettuce	1 minute
Mustard	1-2 minutes	Wood sorrel	1 minute
Nettle leaves	2 minutes	Yarrow	1 minute

> Some inexpensive freezing tools you'll need are:
> - Plastic wrap
> - Large plastic bags
> - Labels or marker pen

To freeze vegetables, steam and drain the vegetable in a sieve; save the liquid for stock if desired. (See Soup Recipes starting on page 198.) Place one-foot-long piece of plastic wrap on the tray. Add 1/2 pieces of steamed vegetable and fold the edges in. Roll package tightly to seal. Any remaining liquid will seep out of package. Place the package in a large plastic bag and label with contents and date of freezing.

Vegetable Wraps

Freezing rolls of vegetables, flowers in blocks.

Freezing wildflowers is a simple matter of freezing whole flowers in solid blocks of ice. A frozen whole rose looks fantastic in a punch bowl. Use an ice cube with a rose inside instead of plain ice. As long as you use a container that will take freezing, you can create a bunch of frozen roses that looks like a crystal dream. My favorite luncheon surprise is layering a selection of flowers in different colors in a large, round bowl. I put in the bottom layer, add a little water, and freeze. A day later, I add another layer, add water, and freeze. And I repeat this until the display is to my liking.

To freeze whole roses, for instance, put a few inches of water in your containers, so that the flower does not float. Keep the pitcher of water handy so that you can reach it easily with one hand. Hold the stems in your other hand, and arrange the colors and shapes to suit you. (I cut the stems to uniform lengths first, making the job of holding them easier.) Then pour some water slowly around the sides, holding down the stems. Put on a lid or plate to hold the arrangement and freeze.

Whole heads of roses opened moderately and without the stems can be layered easily. The more roses in the container, the better. They won't remain in exactly the same position as you put them once you pour in the water, but if they don't float around a lot, the layers will be moderately undisturbed.

Pansies are relatively simple to freeze. Layer them in a sandwich bag and add a little water, then seal the bags. To freeze Queen Anne's lace flowers, spread plastic wrap or wax paper on a flat surface. Place the flowers face down on the wrap and roll up the wrap like a rug. If you have difficulty rolling a single piece of wrap with flowers, place cover sheet of wrap over the flowers and roll. Freeze the roll. Whenever you want some flowers, unroll the wrap take what you need, re-roll and return to the freezer. Queen Anne's lace flowers will turn brown when frozen, but they retain their delicious carrot taste and apparently their nutrition as well.

Freezing Queen Anne's Lace flowers.

Besides vegetables and fruits, you can successfully freeze vegetable stock and leftover fruits. Save all liquid from steaming or boiling vegetables or from leftover tea. The stock is best stored in freezer containers, ice cube trays, or heavy duty plastic bags. Soup stock can be frozen in layers until the container is full.

To freeze fruits, possibly for a mixed juice, layer the fruits in a large container as they are available. You might have a cup of raspberries—not quite enough to make jelly. Freeze in a large plastic container. Are there blueberries left over from making a pie? Add these to the fruit cache. When you have filled your container, you'll have the makings for a mixed fruit jelly or juice.

Decorations Can Be Edible Too

A blustery winter wind whistles outside the door, and my spirit needs a cup of hot rose tea—rose tea is filled with vitamin C, for muscle repair and vitality, nerve restoration, and a general feeling of well-being. From the rose-stem wreath on the north wall of my home, I break off a bunch of twigs, a few leaves, and one small dried flower. After rinsing and placing them in a tea pot, I cover the rose stems, leaves, and buds with boiling water. I let it steep for five minutes savoring the bouquet, and then serve myself tea from my old china teapot. Nature's bounty!

I had ventured out to a spot in the woods where wild roses grew. There they were, inviting me to make lush wreaths from the green shoots.

I took my loppers, and a big bag or length of twine, and began clipping off a few vines. I wound two vines together to make a circle, then added string to the overlapping ends to secure the wreath. I twisted more vines on the base, poking the ends through. When I finished, my decorative rose wreath was hung on the wall to provide a winter's supply of tea as well as to be a feast for my eyes and nose.

My home abounds with wild food wreaths. Circular, heart-shaped, sprays—all wreaths are either leaves or flowers for teas, grains for bread flour, vines or leaves for vegetables, and twig spices for herbal teas. I decorate every nook and cranny with food wreaths.

Wreaths made from wild grains are not only beautiful but practical. Since prehistoric days, people have twisted vines into coils and wreaths, combining craft with foraging. The Chippewa nation still does this today with wild rice hoops or wreaths.

What plants make the best food wreaths? Here's a list.

Food Wreaths	**Grain Wreaths**	**Tea Wreaths**
• Amaranth	• Bulrush	• Balsam fir
• Aster	• Crabgrass	• Birch
• Chicory	• Phragmities	• Blackberry
• Cholla	• Spice wreaths	• Maple
• Clover	• Catnip	• Meadowsweet
• Daisy	• Chamomile	• Pine
• Dock	• Clover	• Raspberry
• Fireweed	• Mints	• Rose
• Grape	• Strawberry	• Sumac
• Lamb's quarters	• Thyme	• Willow
• Malva neglecta	• Wintergreen	
• Queen Anne's Lace	• Yarrow	
• Rose		

Food Wreaths (from left to right, top to bottom): Yarrow tea, Grape vine with tendrils for food, Raspberry canes for food and tea, Rosemary sprig for spice, Spice rope of thyme, mint & chamomile, (2nd row) Balsam fir & meadowsweet flowers for tea and food, Balsam sprig, Grasses, oats & dock seeds for food and flour, and Rose hoop for tea.

If you grow your own roses, treat them with a spray of mild soap and water instead of pesticides and herbicides to ensure their future edibility. Be sure your bush is at least 100 feet away from any busy street. And be aware of your neighbor's habits with sprays and wait a season to eat transplanted nursery stock.

When the time comes to buy your holiday tree, make sure it is organic. Christmas trees are often sprayed for insects and sometimes sprayed green to darken the needles. Find an organic dealer or cut your own tree. When the holidays are over and your tree is dry, spread a sheet around the base and shake off the dry needles. Place needles in a large oven container (a roaster, for example). Heat the oven to 300 degrees and then turn it off. Place the roaster with the pine needles in the oven and remove when the oven has cooled. Do not try to bake the needles! The resin is combustible. Put the completely dry needles into glass containers and enjoy five years more of your Christmas tree as tea.

The trimmings from your tree also make fantastic tea wreaths. Collect them before they are completely dry, rinsing and swishing any dust off the needles. Shape into wreaths and hang in your home for a source of fresh tea for weeks.

Notes

Left: garden sorrel   Right: wintergreen

Part II: Field Guide to Wild Foods
Edible Plants

This Field Guide section provides illustrations and descriptions of the fifty-nine foods discussed in this book. It is not intended to substitute for a comprehensive field guide to wildflowers, trees, or plants. Rather, it should be used in conjunction with three well-illustrated, comprehensive guides to the plant life of your region. Also please note that plant sizes vary with environment, and measurements given here are intended as general guidelines only.

The material that follows is organized as following, information where applicable:

Common name.
Scientific (Latin) name of species, or genus, only if the entry describes a group of related plants.
Family to which the plant belongs.
Other Names: Other common names by which the plant is known; also, types are described if the entry deals with a group of plant rather than a species.
History: Brief background information on the origin of the plant and a notation if used by Native Americans.
Habitat: General growing requirements—where the plant is likely to be found.
Characteristics: Whether the plant is a tree, shrub, or herbaceous plant; whether annual, biennial, or perennial; major physical characteristics to help you identify it in the field.
Primary Uses: Whether used for food, medicine, or as a cosmetic; some mention of the uses the plant has been put to (not meant to be comprehensive).
Nutritional Value: When applicable, a general indication of the plant's value to human nutrition. See also the Nutritional Value tables at the end of Part II for further information on the nutritional value of these plants.
Medicinal Value: When applicable, information on medicinal uses of the plant. See the Appendix for explanations of these medical terms.
Cosmetic Value: When applicable, an indication of how the plant is used as external treatment or otherwise.
Collection and Storage: Tips on harvesting and preparation for immediate use or long-term storage.
Caution: There are some wild plants that can be mistaken for these edible plants. See the section on Poisonous Look-Alikes,

Part V. Note also that entries in the Field Guide include occasional cautions or warning notices about collecting and handling some plants. Be sure to heed these cautions.

Comments: Personal remarks and additional information on the plant and its uses for us all.

Foreground: fiddleheads

Legend to the Uses of Wild Foods

These symbols appear in the description of each wild food and indicate the main uses of that plant at a glance

Beverage	Salad
Bread	Soup
Casserole	Sweets
Cereal	Tea
Flour	Toothpaste
Jam	Vegetable
Medicine	Wine
Pickles	

These symbols are used later in the poisonous look-alike section.

Edible   Poisonous

51

Aloe Vera

Aloe perfoliata vera
Family: *Liliaceae*

Other Names Medicinal aloe, Barbados aloe.
History: Probably native to Mediterranean. Used by Native Americans.
Habitat: Temperate zones, desert, arid regions.
Characteristics: Bushy succulent. Green fleshy leaves, called spikes, are filled with clear gelatinous liquid. Mature plant has spikes up to 3 feet high with dense, arrow-shaped clusters of yellow or orange flowers.
Primary Uses: Flowers, edible raw. Medicinal, cosmetic. In tea as an antispasmodic, antihistamine (asthma, colds, congestion), anodyne, or as tranquilizer or diuretic. Crushed leaves yield gel which is applied as poultice for burns or sores, promoting growth of new tissue.
Medicinal Value: An antibiotic, astringent (to accelerate wound healing), coagulating agent, biochemical bandage, pain inhibitor, growth stimulator, "wound hormone", demulcent, antihistamine, mild laxative, burn healer.
Cosmetic Value: Skin conditioner, astringent. Also aloe has bacteria- inhibiting qualities that make it an excellent underarm deodorant.
Collection and Storage: Harvest in summer; buds, flowers, young leaves. For external use, slice leaf in half and place directly where needed. For internal use, remove gel and place in water. Refrigerate or freeze. Freezes well. Where aloe grows naturally, both young and old leaves are thick. Since old leaves are tough and the gel is bitter, harvest the new, thinner leaves.
Caution: Do not use as bulk food. Use sparingly for medicinal purposes only.

> **Linda Says-** Aloe vera is the most important native plant in the West. Preparing aloe is fun. I take a leaf, turn the flat side down, and lay it on a flat surface. The skin comes off easily by slicing under the skin from tip to larger end. The gel scrapes off quickly with a spoon, Then I scoop the gel into a bowl, whip it thoroughly with an egg beater, and put it into an ice cube tray to freeze. I use one cube to a 6-ounce glass of orange juice for a tonic. One cube also works wonders for immediate application on sunburns.

While aloe is a Western plant, mullein in the East has many of the same properties.

Different Plants, Similar Uses

ALOE VERA	MULLEIN
Antihistamine	Antihistamine
Burn healer, pain reliever	Burn healer, pain reliever
Growth stimulator	Growth stimulator, reduces swelling
Wound hormone	Wound healer-promotes quick healing
Pain inhibitor	Pain inhibitor
Astringent, accelerated wound healing	Whole leaf bandage, antiseptic
Demulcent	Antispasmodic
Coagulating agent	Tranquilizer, biochemical bandage
Antibiotic	Mild sedative
Deodorant, inhibitor of bacteria	
Mild laxative	Anadyne, a medicine that relieves pain
Cosmetic, skin softener	Cosmetic, skin softener

Amaranth

Amaranthus retroflexus
Amaranth family, *Amaranthaceae*

Other Names: Green amaranth, Pigweed, red root, carelessweed, choohugia (Pima name).

History: Native to tropical America; naturalized throughout the world. A staple food of the Zapotec Indians of Mexico. (Sturtevant 1972) Coastal Algonquins collected amaranth for a vegetable and used its ashes as salt. The plant is naturalized in Asian countries. It is cultivated in tropical Africa and Jamaica as a potherb. Seeds are used for flour in India and Nepal. Amaranth yields 8 ounces of seed from plants covering 1 square yard of ground.

Habitat: Cultivated soil.

Common Amaranth (Western U.S.)

Characteristics: Annual herb. Averages 2 foot, but may reach 6 feet or more. With bristly seed heads, clustered on multi-branch stems from a central stalk. Flower seeds are black and shiny; leaves are smooth and veined with slightly toothed margins. Eastern amaranth seed heads or "flowers" are denser and shorter than the Western species. Western varieties may have long, spindly, bristly seed heads, and white seed pods with black seeds inside.

Primary Uses: Culinary, cosmetic. Use leaves and stems like spinach, eaten raw, steamed, sautéed, cooking liquid is drunk. Leaves are also dried and ground for flour. They are used in soups and stews. Seeds are used raw or dried for baked goods, cereal, mush.

Common Amaranth (Eastern U.S.)

Nutritional Value: High in vegetable protein. High calcium and vitamin E.
Cosmetic Value: Astringent, wrinkle cream.
Collection and Storage: Use entire plant. Harvest lower leaves and branches in summer as vegetables. Refrigerate or freeze; dry. Wait until plant is full grown for large seed heads. Amaranth seeds are easily collected in autumn by tapping the seed head over a bowl, even in summer on the desert.

> **Linda Says-** The first time I saw amaranth I couldn't believe my eyes! Tall, wondrous plants heavy with seeds bordering pasture and barn areas. There were hundreds of these "weeds" some bent over from the weight of their seeds. In less than 20 minutes I bundled enough amaranth for a week's supply of fresh vegetable and almost a winter's supply of flour. It did not take long to discover the wonders of amaranth in brownies, as a cooked green, or as a gruel.

Arrowhead

Sagittaria species, *S. chinensis*
Water Plantain Family, *Alismataceae*

Other Names: Duck potato, wild potato, Wapatoo (Indian).
History: Native to North America. Lewis and Clark used Arrowhead for food, taught by Native Americans. The explorers wrote, "(they) can be ground fine. A flair for pudding, cakes, etc. They are nearly equal to Irish potatoes, and are a bread substitute." Chinese sell it in marketplaces for food.

Habitat: Widely distributed. Marshes, wet bogs, mud flats.
Characteristics: Aquatic perennial with tubers. Grows in swamp water to a height of about 10 inches. Arrowhead-shaped leaves and filmy white-petal flowers. Roots have walnut-size tubers.
Primary Uses: Culinary. Root corms: Used as potatoes, sliced or whole, boiled, sautéed, or raw. Strung and dried then ground for flour.

Nutritional Value: Easily digestible; a nutritious food, especially for convalescents.
Collection and Storage: When harvesting arrowhead tubers, use a pitchfork and lift gently. The tubers tend to snap off in the mud and the pitchfork creates less tension. Freezes and dries well. Scrub the small tubers thoroughly and string with a needle and thread. Hang the strings to dry, then store in glass.
Caution: If water purity is in doubt, use purification tablet and soak tubers in water purifying solution.
Comments: There aren't too many arrowheads in my area of northern New York, but southern New York and areas of New Jersey have immense quantities in marshes and wetlands; they are also in wetlands in the West.

> ***Linda Says-*** My marshes do have a few "secret" areas for Arrowheads. Misty, boggy, muddy, slimy, and mossy, the marsh's loam logs rot away to the richest, blackest soil I have ever seen. Mosquitoes are an inch long where the arrowheads poke up above the mud! I pray silently, dip my pitchfork under the plant, and lift slowly. I used a needle and thread to string the tubers in the cabin, drying them for the winter. Easily reconstituted in water, the tubers make delicious soups and stews.

Aster

Aster nemoralis
Composite Family, *Asteraceae*

Other Names: Blue, pink, lavender, white & yellow aster.
History: Native to North America. Staple of the Iroquois Indians in the Northeast; food of tribes in the Midwest and West. Used in sweat lodges by Objibwa Indians to revive consciousness by putting on the hot rocks. Used as a food in South America.
Habitat: Fields and meadows, roadsides.
Characteristics: Perennial herb. Many varieties. Height varies, average 24 inches; may reach 40 inches. Numerous extensions jutting with multi-petaled flowers from a central stem. Petals are layered in flat, 1/2-inch-long sections. Colors vary from white to pink, yellow, blue, and purple.
Primary Uses: Culinary, Medicinal. Leaves eaten raw, steamed, boiled, sautéed, cooking liquid is drunk. Flowers stewed or steamed whole; chopped fine and stewed or steamed. Entire plant is dried, then stripped and ground for flour. Sprinkled on dishes, especially tomato-based dishes such as lasagna. The pith of the aster stem is edible, even in mid-winter. As emergency food, eat raw or cook in soup.
Medicinal Value: Dried stems and flowers are used as a wash for rheumatism; also excellent blood tonic tea. Roots of dwarf purple aster used for diarrhea control.
Collection and Storage: Snip the long stems at ground level and bundle together with string. I carry 2-foot lengths of string around my shoulders, and harvest the flowers until I have a bundle 10 to 15 stems thick. Stop bundling stems together when the flowers are too tight to be airy for drying. Take only a few plants from each

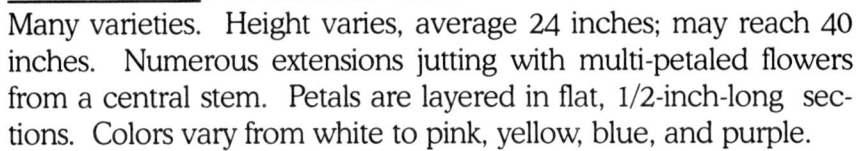

area (or 1 out of 4). Hang bundles upside down on a clothesline to dry. When dry, strip down onto sheet, crumble, and store. Also, leaves can be frozen.
Caution: Do not eat domestic asters.

> ***Linda Says-*** Nature's abundance is particularly clear to me when I harvest the Iroquois staple, the wild aster. There are 250 varieties of wild aster in the United States: blue, lavender, yellow, and white prevail in the Adirondacks; blue, lavender and yellow in Western states. Field may have flowers as far as the eye can see. In less then 15 minutes I have enough to make a winter's supply of wild aster flour. I grind this bounty into flour for baking from dried storage food. The flour is particularly strong, nutritious. I recommend adding a small amount to regular recipes. This Iroquois staple remains one of my favorites.

Balsam Fir

Abies balsamea
Pine Family, *Pinaceae*

Other Names: Christmas tree.
History: Native to North America. Soft inner bark was used as emergency food at Valley Forge. Native American usage.
Habitat: Grows in cool, moist, acidic soil along roadsides, woods, or lawns and in softwood forests.
Characteristics: Evergreen tree. Has small, short needles growing alternately from under stem, 1 to 1-1/2 inches long. Grows to height of 40 to 60 feet.
Primary Uses: Culinary, medicinal, cosmetic. Bark is chewed, ground or dried for flour. Sap is chewed as gum. Branches and

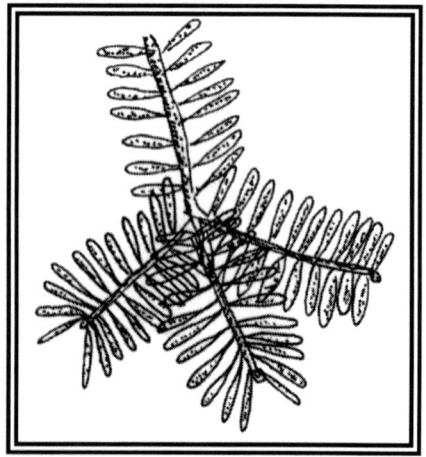

twigs are stripped and ground for flour. Needles are used raw for food; steeped raw or dried for tea. Used in wine. Ground raw or dried for emergency flour. Dried needles are used to fill balsam pillows and sachets. They lend a fresh scent to drawers and closets.
<u>Nutritional Value</u>: Good source of vitamin C.
<u>Medicinal Value</u>: Roots used as a poultice to soothe syphilitic sores and gonorrhea. Sap (pitch) is a "glue" for cuts, scratches, bites, and sores. Dried sap is chewed for colds, bronchitis, influenza, pneumonia, upper respiratory diseases.
<u>Cosmetic Value</u>: Oil used for flavoring, scents, perfumes, soaps.
<u>Collection and Storage</u>: One branch yields a winter's supply. Balsam fir tips are a lighter green, indicating newer growth. I snip off needles here and there, filling a bag with dozens of twigs. Needles dry easily and quickly indoors in a sunny spot. For tea, simmer 1 cup needles until water is dark green. You may reuse needles or bark several times, adding water and simmering. Poke sap blister and gather liquid for emergency glue; store in glass if possible.

Linda Says- Rather than throw away your Christmas tree, use it for nutrition as well as a room refresher. For a wonderful aroma throughout the winter, simmer needles in water until they reach a thick, gummy state. Use twigs and parts of older branches as a fresh-smelling incense. To dry sap for medicine, make small flat cakes of sap and dry in a thin layer at the back of a wood stove or above the pilot light of a gas stove.

Birch

Betula species
Birch Family, *Betulaceae*

Other Names: White birch, paper birch, yellow birch, golden tree.
History: Native to North America. Scandinavians boiled, baked and added birch sawdust to their breads. Loggers say that their white birch sawdust used to be taken to American bread companies. Native Americans used birch dust for tea, and the strong inner bark was fashioned into many crafts. The Cree Indians folded birch bark and bit patterns into it, then unfolded it. Hundreds of Confederate soldiers were saved during their retreat to Monterey, Virginia, when they used birch bark as food[1] (Harris 1973).
Habitat: Woods, roadsides.
Characteristics: Deciduous tree. Alternate simple saw-toothed leaves. Papery bark of white and yellow birch peels in curls.
Primary Uses: Birch is an excellent cabinet-making wood; it makes strong hardwood furniture. Culinary, medicinal, cosmetic. Inner bark, sap, twigs, buds, and young leaves eaten raw as emergency food; dried and ground for flour. Sap is drunk raw for nutritious liquid. Twigs are used to make wines, dried for tea or crispy treat. Buds eaten raw. Young leaves steamed, sautéed, cooking liquid drunk.
Nutritional Value: High in minerals, calcium and phosphorous. High in potassium and beta carotene.

---

[1] Ben Charles Harris, p. 444

Medicinal Value: Buds are a tonic, since birch bark contains salix, or acetylsalicylic acid. A strong tea made from the bark has been used as an aspirin substitute.

Cosmetic Value: Birch bark used for skin problems, eczema, skin eruptions, pimples, dandruff, and scalp disorders. Dried, powdered leaves used for chaffed skin; birch oil for hair tonic or body oil.

Collection and Storage: Use the bark of downed trees, or follow a logging trail and gather the sheets of bark left behind. You can also peel the bark off firewood in the spring. Strips of birch bark make a quick, sure fire. Four 5 1/2 inch strips = 2 cups fine flour (a winter's supply). One branch approximately 4 feet long = winter's supply for 4 people.

Tap the trunks in spring (April-June) like maple trees. One tree yields an average of 1 gallon of syrup in a few hours. Twigs, inner bark, and sap can be used all year. Buds and twigs have a wintergreen taste. The sap can be boiled down and used as a syrup or in birch beer. Boil down birch syrup further to make birch molasses. Store dried and ground inner bark in glass as an emergency food. The inner bark makes a delicious mineral tea; carry a chunk with you when doing hard physical work. The inner bark dries to a breakable sheet ready to be used as tea; store in glass containers.

Caution: People allergic to aspirin may have a reaction to salix wild foods such as birch.

Caution: Peeling the bark of a living birch will kill the tree. This is called girdling a tree. The bark is literally the skin of the tree, and without it the birch will die.

Linda Says- Birch bark is one of my most precious possessions. I use it in every season for tinder and generally keep sheets of it in quantity. In the middle of one winter I received an order for birch-bark baskets. Since I had no bark on hand, I got permission to go up a logging trail with Ken and Todd. What a day-20 degrees below zero, with 4 feet of snow and no snowshoes! As I peeled bark off stumps of white birch with my knife and hatchet, Todd played in the drifts. That night my back ached in remembrance of the work, but I had enough bark for hundreds of baskets, with plenty left over for Christmas cards and for starting fires.

Blackberry

Rubus villosus and other bramble berry species
Rose Family, *Rosaceae*

Other Names: Creeping blackberry, cloudberry, wineberry, dewberry.

History: Native to North America and elsewhere. Used as a fruit in all countries where found.

Habitat: Roadsides, fields, meadows.

Characteristics: Biennial trailing stems. Bush varies in size, up to 5 or 6 feet tall. Saw-toothed leaves in groups of 3 to 5; thorns on angular, arching stem; fruit pebbly and black when ripe. Canes produce fruit on second year's growth. Fruit does not separate from receptacle when picked.

Primary Uses: Culinary, medicinal, cosmetic. Young stems and shoots eaten raw or cooked; used fresh or dried for tea. Fruit is eaten raw, and cooked for jellies, jams, syrups; also juices and wines. Flowers entirely edible raw. Young leaves are edible as a cooked vegetable in the spring.

Nutritional Value: Young stems and fruit are good source of vitamin C.

Medicinal Value: Bark of stem helpful for intestinal problems. Tea made from blackberry root helps control diarrhea. This tea is also an astringent, a mouthwash for mouth sores. As a strong tea or eaten raw, all parts of this plant can be used to relieve diarrhea or as a blood cleanser.

Cosmetic Value: Leaves are good for facials, masks, lotions, and astringents. In an extract, clears blemishes, eases scalp itches, and heals scales.

Collection and Storage: Tie a lightweight receptacle around your waist and use both hands to pick berries. Hand-pick young leaves in early summer.

Picking Blackberries in Adirondack Bear Country

The dirt road next to the shanty led down to an old log bridge. The bridge spanned the Cedar River, and there was nothing but private wilderness for several miles. I crossed the river and climbed the wooded path for one mile. In another half-mile, I came to the largest blackberry patch I had ever seen. I was equipped with lightweight containers for the berries and fly repellent made of olive oil and pennyroyal.

I whistled a tune as I approached the patch, wary of the bears known to be in the area. I knew if I blended in too much, the bears might resent it. This philosophy had worked for me before-either I was allowed into their domain or I was not. Then I became aware of a very large body up ahead. I could hear and smell the bear shuffling slowly through the woods. I waited. Testing the wind for the bear smell and listening with my ears as I whistled, I knew the bear shuffled away.

I picked blackberries with both hands, dumping them into a plastic bottle and then into a large bucket. I worked my way around the path from left to right. Experience told me that I needed two hours to obtain eight quarts of juicy blackberries, and this particular day I picked eight quarts and then left the patch to the bears again.

When I worked with my son, Todd, we had our two coffee cans filled in moments and proceeded to dye our sweatshirts blue-black with the weight of the juicy fruit. With over a gallon of ripe blackberries we made jelly that night over an evening campfire. Breakfast was a treat!

Blueberry

Vaccinium myrtillus
Heath Family, *Ericaceae*

Other Names: Bilberry.
History: Native to North America. Used by Native American tribes.
Habitat: Acid soil, fields, roadsides, waste areas, bogs, marshes, woods.
Characteristics: Low spreading plant or woody shrub. Many varieties, from ground-hugging plants to high bushes. Fruit is in clusters in groups of 2 to 10. (Huckleberry often grows near blueberry; edible fruit looks similar, but huckleberry has stony pit and tart taste.) Wild blueberry looks like bilberry: both are edible, although true bilberry is not very tasty.
Primary Uses: Culinary, Medicinal. Fruit is eaten raw or cooked for jellies, jams, syrups; also wines.
Nutritional Value: High source of Vitamin C.
Medicinal Value: Blueberry leaves used medicinally in a tea. Because they contain myrtillin, an antispasmodic and relaxant, blueberries were used as a drink during childbirth to relax the mother. Leaves, fruit, and twigs can be made into an infusion and drunk for kidney problems. Dried berries are chewed slowly and thoroughly to aid control of diarrhea. Leaves and twigs can be used separately or together as teas, tisanes, and throat gargle.
Collection and Storage: Blueberries are best picked by first pulling sections with both hands rather roughly and letting berries fall on to a sheet spread around the bushes. After rough-picking a few bushes, rustle the branches for the many berries caught between the leaves. Gently gather up the sheet and pour the berries on a tray.
- 4 cups fresh berries = 8 jars jelly
- 4 cups fresh berries = 2 cups dried berries
- 4 cups fresh berries = 1 quart frozen berries

Twigs can be dried or dug up under the snow. Berries are placed in cold water. The immature berries and leaves float to the top for easy removal. To dry, spread on a tray; cover with a screen or cloth. For storage, berries can be dried until they rattle, or frozen.
Caution: Positively identify; crush fruit for characteristic blueber-

ry smell. Blueberries have a slight frill and an opening on the side away from the stem. A species of buckthorn, which is poisonous, has blue berries without a crown or frills.

***Linda Says*-** The fields behind the old McConnel place were filled with ripe berries, and became my favorite place to pick. I'd place a sheet around the bushes, and use both hands for picking. With a sheet you can become lightning fast, but it is still a challenge to pick enough berries for a pie before the blackflies find you. I usually made smudges before the clouds of flies found me if I wanted to take my time picking. Then I'd dry blueberries on screens covered with a cheese cloth. Those berries in my cereal on a winter's morning made it seem like spring.

Bulrush

Scirpus validus and other sedge varieties
Sedge Family, *Juncaceae*

Other Names: Sedge, tule, great bulrush, nupiaskun (Crow Indian.)
History: Found nearly throughout the world. Used as food from Moses' time including use by all Native Americans. The Crow made mats by sewing together bulrushes with basswood string and a bone needle.
Habitat: Wet areas, drainage spots, bogs, swamps; alongside waterways and ocean bays.

Characteristics: Grasslike herb. Many varieties with height to 24 inches or more. Stems smooth, round, with pith; head of flower spike full of seeds and pollen.

Primary Uses: Culinary, medicinal. Roots cooked as potato. Early shoots eaten raw, steamed, boiled, sautéed; cooking liquid drunk. Flower clusters dried and ground for flour. Bulrush flour is sweet and nutty—it has the most branlike taste of all the grains. Pollen used as flour. Seeds ground and roasted.

Nutritional Value: Highly nutritious, roots rich in complex carbohydrates.

Medicinal Value: A plaster or poultice for aches and pains.

Collection and Storage:
- 6-8 bulrush plants = up to 1 cup leaves
- 1 cup leaves = 1/2 cup bulrush flour
- 1 teaspoon powdered bulrush leaves = 1 cup tea

See page 34 for collection details.

Dry the bulrushes whole, bundle the long stems and hang them upside down in a warm, airy spot out of the sunshine. To ensure dryness, dry in a 300° F. oven for 5 minutes, then turn oven off and wait until cool. Follow this drying procedure even if using the bulrush strictly for decorative purposes.

Linda Says- I have spent many wonderful mornings watching dragonflies and frogs playing in the swamp while I gathered bulrushes. Swamping is not only fun, but profitable-I've fed many people on bulrush bread.

Insects love to lay tiny eggs in bulrush tops, so even if I am using the bulrushes for decoration, I leave them outside for 24 hours so the bugs can leave. After drying, I use them in lovely arrangements or grind them to a fine, delicious flour for baking.

Burdock

Arctium lappa
Sunflower Family, *Asteraceae*

Other Names: Beggar's buttons, burs.
History: Native to Europe and Asia; widely naturalized in North America.
Habitat: Fields, roads, most waste areas.
Characteristics: Bush like, large green leaves over a foot wide, and many veined. Tubular, purple or white flowers followed by spherical burrs, 3/4" diameter. In the fall, burs dry to brown, sticky seed pods.
Primary Uses: Culinary, medicinal. Roots are boiled; dried and powdered for nutrient additive. Stem pith is baked, boiled; eaten as pickle, candy. Early leaves are eaten raw, steamed, boiled, sautéed; cooking liquid drunk. Large leaves are used as trays and receptacles for small plants. Burs when young and still pliable are steamed and eaten; when mature, broken in half and white pith eaten raw.
1 large leaf holds 2 cups green burs
1 8-inch root = enough raw root for 2 or 3 people
Nutritional Value: High in vegetable protein. Contains inulin, the chemical source of insulin.
Medicinal Value: Tea used to heal infections. Root is a diaphoretic; as wash infusion, for poultice, bruises, sore or tired feet.
Cosmetic Value: Seeds for skin problems, eczema, cleansing. For pet's fleas, simmer a small handful of brown burs, strain, and cool. Wash your pets with the decoction. Kills fleas quickly!
Collection and Storage: When you can, harvest the young roots before the flower stalk develops (usually the second year of

growth). When digging roots, slide the shovel thoroughly down one side, then another, until all 4 sides have been deeply cut. Keeping the shovel in one side, pry up the root without breaking it off. You will feel the pop when the root releases from the earth. If the tip remains, the plant will grow again. Steam and wrap burs individually to freeze. Slice and fry or simmer and then freeze roots. Gather burdock in any season. Similar in size to rhubarb leaves, the gigantic burdock leaf forms its own carrier. In spring, place tiny new burs on a leaf, roll the leaf, tuck in the ends, and put the leaf in your pocket. In winter the burs are brown. I take a few brown burs, stroke the bushes, and gather a ball in no time. I stick the burs together in a large ball and carry them on my shoulder, freeing both hands for further picking.

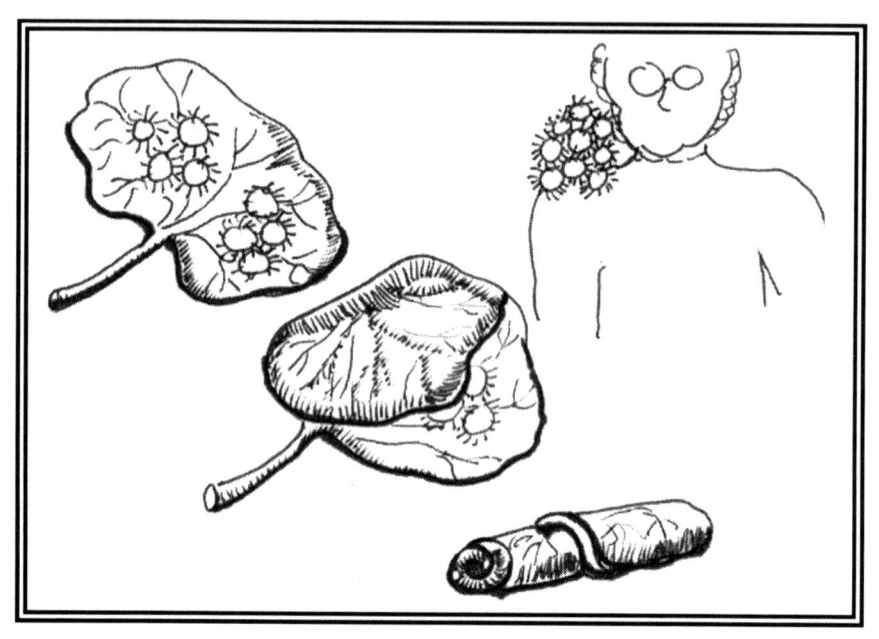

Carrying Burs

Linda Says- When harvesting, cut over the joint of the root where the stem begins. Scrub the root with a brush then peel root until you reach fiber. The fiber will turn from white to brown very quickly; this is normal oxidizing. You can peel the brown off again when serving, but I just steam the burdock pieces and freeze for a winter's meal. Drying the peels makes a crunchy, chewy root food.

Cattail

Typha latifolia
Cattail Family, *Typhacerae*

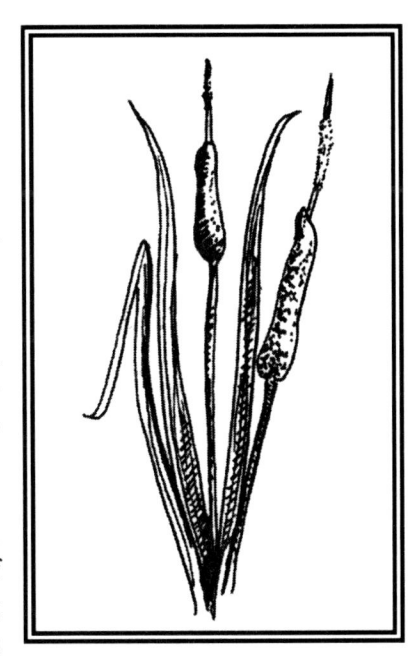

Other Names: Supermarket of the swamp, punk, upakiotipa (Crow Indian).
History: Native to North America, Europe, Asia. Cattail has been used for food throughout recorded history by people in all countries where it is found.
Habitat: Bogs, swamps, wet areas.
Characteristics: Perennial herb. Grows in wet areas to a height of 10 feet or more. Stalks have hot dog-shape heads, pollen flag in early spring. Very tall slender leaves with 1 vein.
Primary Uses: Culinary, medicinal. Roots are dried and ground for flour. Early shoots are eaten raw. Stem pith is eaten raw, boiled, and pickled. Early green heads are eaten raw, cut and cooked as ear of corn. Early brown heads are ground for flour. Pollen is used as nutrient additive. Leaves are used for basket weaving, mats, other crafts.
Nutritional Value: Plant holds about 30 percent complex carbohydrates; highly nutritious.
Medicinal Value: Flower heads used in tea for diarrhea control.
Collection and Storage: Pollen collected in spring; shake into a paper bag. Stems harvested before the cattail flowers, whenever possible. Roots dug from winter to early spring. Gather fluff from mature heads for excellent insulation or stuffing for jackets. The fluff floats and is waterproof, and serves as excellent tinder and torch. Can be used as cotton.
Caution: If water purity is in doubt, use purification tablet and soak plant in solution.
Caution: Pollen fluff may cause skin to break out in hives.

Cattailing Seasons

The morning is misty and cool. The Adirondack. ice has finally left the swamp and the crystals have melted along the edges of the mossy mats. The cattail shoots will soon be through. Water is beginning to show the spring signs of life. I saw bugs skating yesterday and heard a frog peep. Steam is rising from the invisible shield above the water. The old, dried stalks of cattails, left after winter's fury, lay bent and crooked, disheveled. The matted fuzz of the old tails looks pathetic as the new growth heads toward the surface of the shield.

There are blades-green shoots-poking up from the swamp. I take my trusty shears, long boots, plastic garbage bag, 4-foot board, and towel (should I fall in the boggy mud), to gather cattail shoots to can for winter. Laying the board from mat to mat, I gingerly move the length, clipping the shoots and dragging my filling bag behind me. Glancing back into the swamp, I can tell where I have harvested. The shoots will come again, reaching toward the sun of summer.

Later in spring, green plumes meet my eyes. The heads are like corn on the cob. I can taste and smell the meaty, nutritious flesh. My board goes down gingerly now. The swamp is dotted with thousands of green spikes. Gathering carefully, I pick only cattails with a full cob. Making piles of 8 to 10 stems, I bunch them in a pattern behind me, then I wrap each bundle with the leaves, winding them round.

Though the water is much warmer now, steam rises as dozens of insects zoom through. Mosquitoes glide by, legs tucked up underneath, looking like tiny planes with their landing gear drawn for a landing. Frogs and peepers croak and splash off their pads. I have cut the largest harvest ever. Now winding my way back I pick 10, 20, 30 tails at once. I'm coming back to camp with a supply of cobs to savor.

Summer is here and the murky mud is tough going. The cattails have filled the swamp with heavy bushy stems and blades rich with brown cobs. They are ready for harvest. Flour time! Bread time! Pancake time!

The cattail stems have swelled to 1 inch, sometimes 1 1/2 inches thick. Inside lies a creamy white pith that makes delicious pickles. Sweet and highly nutritious, they will supply a winter's fat. The cattail is heaviest now so I gather only a few, counting up to 10. I need 20 to 30 to make a loaf of bread. The mosquitoes are merciless, and the 90 degree August heat is causing the brackish water to make a stench. Under the cattails grow pennyroyal, mint, and other grasses. The harvest is half done for the winter. Another day and I will have seven or eight gallon containers filled with the pure starch piths.

Fall is here and my last trip for the year to the swamp is a sad one. Most mosquitoes have disappeared when the temperature dipped to 50 degrees. Ice crystals will form soon. Soon 40 mile per hour wind will swirl snow. Life will be in a dormant stage for many months now, but the precious cobs will be our insulation, our stuffing for pillows, blankets, and winter jackets. This collection is the easiest. As fast as I cut a cob, the pollen pops out and my bag fills with creamy fluff. The tough brown skin holds some heads together, awaiting a twist by both hands to rip open and fluff out the rich pollen.

The mud has an almost fishy odor, and it is full of algae and frogs. At this last harvest of the year, the rewards come in floods, and the swamp holds my deepest respect. I have reaped its finest fruit-the supermarket plant of the swamp: the aquatic cattail.

Cattails and Cobweb

Chamomile

Matricaria chamomilla
Composite Family, *Asteraceae*

Other Names: Ground apple, Roman chamomile, mayweed, German chamomile, pineapple weed.

History: Some varieties native to North America; many naturalized from Europe. Used throughout Europe. Also cultivated for commercial use, such as in shampoos and teas. Oil is extracted for many products, including hair rinse and insect repellents. The fragrance is called mananilla in Spain. Used especially in Italy, England, and Germany.

Habitat: Roadsides, "people places," walkways, gardens.

Characteristics: Biennial herb. Height average of 3 inches; may reach 12 inches or more. Tight-budded flowers, tiny yellow daisy; sweet, pineapple-like smell.

Primary Uses: Culinary, medicinal, cosmetic. Twigs are dried for crispy treat. Leaves are eaten raw; steamed; sautéed; cooking liquid drunk. Steeped raw or dried for tea. Flowers are eaten raw; steamed; sautéed; cooking liquid drunk. Entire plant is steamed; boiled; used in soups and stews. Steeped raw or dried for tea.

Nutritional Value: High in minerals; calcium, iron, potassium, and niacin.

Medicinal Value: Stomatic, diuretic, chapped skin ointment; root used for toothache.

Cosmetic Value: Compress for eyes, shampoos, rinses for hair luster.

Collection and Storage: Grows near pathways, and a winter's supply can be picked in a few minutes. Shear with scissors.

Linda Says- The smell of chamomile tea means fields, sunshine, and peace for me. I especially like to dry the flowers for tea, but any part of the plant will serve just as well. This lovely plant has more uses than tea, though. I eat the plant raw as a mild tranquilizer. The plant's parts make a delicious addition to salads. Chamomile flowers frozen in ice cubes bring a touch of spring to a drink on a cold winter's day.

To use as a hair rinse, simmer the entire plant -leaves, roots, and flowers-in water. Simmer until water is green, about 15 minutes. Cool the water, strain, and apply. Let this sit on your hair for a while, then rinse again. It will give your hair amazing vitality.

Try seeding chamomile in your garden or lawn and let it go wild.

Chickweed

Stellaria media (common chickweed), *Cerastium* (mouse ear chickweed)
Pink Family, *Caryphyllea*

Other Names: Star flower, chickweed, stitchwort, starwort.

History: *Stellaria* native to Europe and naturalized in North America; *Cerastium* native to North America. Widely distributed; used by Native Americans.

Habitat: Fields, lawns, roadsides, waste places.

Characteristics: Annual or perennial. Flowers petaled, 5 pairs, deeply cleft in half. Tiny and white above long stalk 4 to 6 inches. Leaves ovate and smooth. All types of chickweed are small ground plants with some parts erect but mostly prostrate. Prolific, seen year-round in some areas, even under the snow. Flowers can be found most months of the year. Mouse-ear chickweed has minute hair on all leaves and stems.

<u>Primary Uses</u>: Culinary, medicinal. Whole plant is edible raw. Used as a vegetable or salad green. Also stewed, stir-fried, dried or frozen for later use. Mouse-ear chickweed is cooked to remove hairs.

<u>Nutritional Value</u>: High in calcium, potassium, and ascorbic acid; trace minerals.

<u>Medicinal Value</u>: Soothing effect; upper respiratory; externally for poultices on inflamed skin.

<u>Collection and Storage</u>: The tiny star-shaped flower petals have a tiny notch at the end. Leaves are easily twisted and picked from stems. Gather as you would dandelion leaves (see page 24). Young leaves are best. I have always identified them by this.

Linda Says- A bit of garlic and steaming hot chickweed make a nutritious meal for the forager because it is high in calcium, potassium, ascorbic acid, and trace minerals. I once found myself sitting on a hillside, pulling handfuls of chickweed and eating them with gusto! The most amazing feeling is the delicate leaves of chickweed. When I put my fingers down into the thick mat of tiny nourishing leaves to pull up a handful, my contact with this plant always is one of delicacy and gentleness. A little washing, oil and vinegar and the feeling is easily transferred to my mouth!!!

Chicory

Cichorium intybus
Composite Family, *Asteraceae*

Other Names: Blue sailor, blue daisy, coffee weed, wild succory, Barbe de capuchin, whitloff, ragged sailor, wild bachelor button, blue dandelion, blue daisy.

History: Native to Europe, naturalized throughout much of North America. Used by Native Americans and a staple European food. Each year the United States imports tons of chicory root for use as a coffee substitute, coffee flavor, and coffee extender.

Habitat: Roadsides, fields, dry sandy areas.

Characteristics: Perennial herb. Grows to height of 2 feet but may reach 4 feet. Has deep taproot and milky juice. Red-veined, deep-cleft leaves are thicker than dandelion to touch. Stem has many star-blue ray flowers. When broken, root exudes a white milky sap, as do the stem and leaves. Plant begins in a lettuce-like whorl of leaves on the ground, heads up in a bunch, then shoots a stalk upwards from the center. Roadsides show bunches of deep green, deeply indented leaves; not unlike dandelion, but a bit more uniform in clusters before the stalk appears.

Primary Uses: Culinary and medicinal. Roots are dried and ground to coffee. Leaves are eaten raw in salads and sandwiches. May be steamed, stir-fried, boiled as a vegetable, sautéed; cooking liquid drunk. Stems are entirely edible as are the flowers and seeds.

Nutritional Value: Roots are rich in beta-carotene and niacin as well as carbohydrates.

Medicinal Value: For stomach and kidney disorders.

Collection and Storage: Gather roots in early spring. Roots may be stored in strips for easier grinding (see page 30 for illustration on stripping roots). Store whole roots in paper bags in a warm, dry area for future pulverizing or boiling.

- 1/2 paper grocery bag of dried leaves = 1 quart rough ground flour
- 1/2 paper grocery bag of dried chicory roots = 1 quart ground coffee substitute

      To collect pull up entire plant by grasping stem as close to root as possible and pull straight up. Hack off the root (I use an ax), gather and wash the leaves, and scrub the root with a toothbrush. When the roots are clean, slice lengthwise into strings. These dry easily on a screen or may be strung over a wood stove to dry.

Linda Says- Chicory roots pull up easily from sandy soil. I put my bounty in a rain barrel to keep the entire plant fresh until I can work on it. All summer, Todd mows the lawn where there is chicory. The reddish vein makes it easy to identify the plant. To tell chicory from the common dandelion, just feel both leaves. Chicory leaves are thicker and not shiny. Dandelion is softer and shiny.

Cholla

Opuntia fulgida
Cactus Family, *Cactaceae*

Other Names: Jumping cactus.
History: Native to North America. Used by Hohokam, prehistoric desert people. Indian women used baskets, sticks, and wooden tongs made from saguaro ribs to gather the buds. A firepit was dug in the desert floor, and the buds and joints were placed in a mesquite fire, roasted, and split in two to eat the succulent insides.

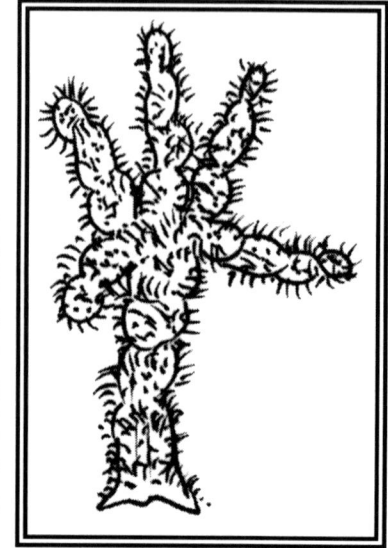

Habitat: Desert.
Characteristics: Treelike cactus with many branches. Cholla cactus has dozens of individual egg-shaped barbarous sections extending from tree-like stems reaching heights of 3 feet or more. Flower is light rose color, fruit is green and smooth.
Primary Uses: Culinary, medicinal, cosmetic. Edible flowers, seeds, fruits, and bud extensions. Fruits are eaten raw, boiled, or baked. Dried for long-term storage. Fruits used in soups, casserole.
Nutritional Value: High in calcium and iron.
Medicinal Value: Gel applied on skin burns.
Cosmetic Value: Gel used as skin softener.
Collection and Storage: Use tongs and paper bags to collect fruit, leaves, and flowers of cholla. Spines and glochids are removed in any of several methods. Indians used flash fire, holding a flame under the burr to remove the glochids, so that they could be opened easily and handled with the fingers. Another method is to place burrs in one paper bag and transfer to another several times. Dry cholla buds on screens in the sun. Cover with cheesecloth if birds pick at them. Dried buds are stored in paper bags until needed. When needed, reconstitute in water about 3 to 4 hours, then boil for one-half hour.

Evelyn Neithammer (1974) found that the easiest way to clean cholla buds is to fill each of 2 saucepans one-third full of clean gravel. The buds are added and the gravel and buds poured from one pan to the other four or five times, or until rid of spines and glochids. (Glochids are minuscule, dense pockets of small barbs which protrude from the pads of prickly pear cactus. They are small but mighty protection against antelope, deer, and cattle of the Western plains.)

Caution: Spines and glochids will penetrate skin with a voracious sting and burning sensation. Do not touch the cactus with bare skin.
Caution: All cholla, prickly pear, and saguaro cactus is "protected plant, by State of Arizona," but it is legal to pick fruits and buds of the species in this field guide for food. The rare crested saguaro is completely protected, so NO fruits or parts may be taken. Be sure to check the regulations in your state.

Teddy Bear Cholla

My first introduction to cholla is a painful, clear memory. My brother, Paul, and I were taking a first tour of the desert together. I was listening to him carefully. "Walk behind me, check your shoes carefully for the cholla burrs, and never step over the bushes here...walk around in the bare spaces." I was trying to listen, but the 100 degree temperature became a problem and I began doing my Adirondack way of walking. I stepped over a bush, but never put my other foot down. It was stuck to my rear with a thousand fire barbs. (The burrs were attached to my sneaker heel, which brushed my rear closely enough to get the burrs attached to my skin.) Screaming, I stood an embarrassed 20 minutes while Paul picked out the hundreds of minute hairs with tweezers. Needless to say, I remember the rules now and thank him for a lesson well learned!

Clover

Trifolium pratense (red clover)
Trefolium repens (white clover)
Legume Family, *Leguminosae*

Other Names: Long stalk, strawberry clover (red clover); common, sweet, short stalk clover (white clover).
History: Naturalized from Europe. With over 300 species, clover has been used by all cultures throughout history. Native Americans used it widely as a vegetable or cure for chest congestion.
Habitat: Roadsides, fields, lawns.
Characteristics: Biennial or perennial herb. Red clover reaches height of 10 inches or more, with hairy stems. Red or purple blossom with oval nectar sections; elongated leaves form trefoil with white vein when mature. White clover reaches height of 2 inches or more. White blossoms have dozens of nectar filled sections; round leaves form trefoil at end of stem.
Primary Uses: Culinary, medicinal, cosmetic. Leaves are eaten raw; steamed; boiled; sautéed; cooking liquid drunk. Dried and

Red Clover

White Clover

ground for flour. Flowers eaten raw; steamed; sautéed; fried; cooking liquid drunk. Dried for tea; used for wines. Seeds crushed for cereal; sprouted. Entire plant is steamed or boiled and used in soups and stews; cooking liquid is drunk.

Nutritional Value: High in calcium, potassium, niacin and vegetable protein.

Medicinal Value: Red clover is used as tea for cough, whooping cough; blood tonic or purifier. Clover syrup used for chest congestion and bronchitis.

Cosmetic Value: Cosmetics, facial creams, rinses, shampoos, wash for pimples, poultices for athlete's foot fungus.

Collection and Storage: Plants are most succulent in spring and early summer. Gathering a winter's supply of clover takes only a few minutes. Clover can be frozen by placing it in a single layer on freezer wrap, folding over 2 sides to hold the clover in place, and freezing. After the clover is frozen, **Sheering Clover** roll the paper to make a compact package, fasten, and label. Dry seed heads separately for an attractive potpourri.

Linda Says - White clover has a particularly different flavor from red clover, with a smaller flower and leaf. The white clover leaves are easily sheared for food as they grow close to the ground in patches. One clover patch 3 feet wide will provide dozens of servings in one season. One summer I conducted an experiment. Shearing a 3-foot patch every morning for 3 weeks, I successfully fed 200 people several different dishes of clover: clover with rice or potatoes, clover casserole, breaded clover heads, clover tea, candied clover heads. The clover flowers went well in spaghetti sauce for meatballs too!

> **A Protection Powerhouse**
>
> Red clover borders almost every road edge. It appears to grow on all lawns, & is found in dozens of varieties throughout the world. When I discovered that the plant leaves were vegetable protein, I was ecstatic. I had been walking into town every two weeks for supplies, including cheese. My biggest nutritional concerns were protein and calcium, & then I read that clover has tremendous amounts of protein. Indeed, five large clover leaves provided the same amount of vegetable protein as 1 ounce of cheese. That did it! I began to put clover leaves between two pieces of whole wheat bread, calling it cheese. Within three days I was eating the leaves raw, saying "I need energy." I would forage freely on a small amount of raw leaves.
>
> A vegetable, a tea, a vegetable flour, clover has become one of my staples. I add clover to rice dishes, simmer clover soup, & steam casseroles with clover as an ingredient.

Crabgrass

Digitaria sanguinalis L.
Grass Family, *Gramineae*

Other Names: Finger grass (Horizontalis, ischaemun (smooth); serotina (creeping); longiflora; simpsoni; filiformis, and more.)
History: Native to North America, Asia, Mediterranean and North Atlantic coast as well as Europe. A drink called Mannagrit is a popular recipe. Used in most areas of the Mediterranean as flour for pita cake.
Habitat: Lawns, roadsides, waste places, edges of fields. The large crabgrass prefers moist soil, streams, ditch banks.
Characteristics: Grasslike, many varieties reaching heights from 3 to 24 inches or more. Summer annual and perennial, the bract or seed tips may vary in amounts and thickness, as well as smooth seeds or heavy bracts.

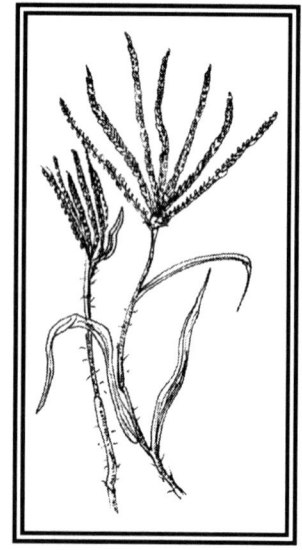

Primary Uses: Seed bracts as well as leaves are dried and ground to flour. Used raw, steamed, boiled, sautéed, and cooking liquid drunk. Pollen used as flour; seeds are sautéed; stir-fried, or ground and used as flour.
Nutritional Value: Highly nutritious; roots rich in complex carbs.
Medicinal Value: Used as a poultice for aches and pains.
Collection and Storage: Shear bracts and leaves, as if mowing lawn. Dry thoroughly, store whole in glass. Grind to flour when needed.

Linda Says- I'll never forget the feeling when I realized that a 2-foot section of crabgrass would yield a batch of muffins every 2 weeks or so. When I read that the Bedouins pulverized crabgrass to a rough flour and made a cake the same as I did, I knew then that my cakes were a universal recipe.

Daisy

Chrysanthemum leucanthemum
Composite Family, *Asteraceae*

Other Names: She loves me, she loves me not; ox-eye daisy.
History: Native to Europe and Asia, widely naturalized throughout North America.
Habitat: Fields, roadsides, lawn, waste places.
Characteristics: Biennial herbs. White petaled flower about a long, erect stem 18 to 20 inches tall. Flower center has a yellow pebbly disk of seeds. Flower ranges from 1 1/2 to 2 inches in circumference. Petal number varies. Leaves are dark green and

irregularly lobed in a scalloped design. Smooth and succulent, they form a thick basal rosette the first year of growth, sending the stem up with flower the 2nd year.

Primary Uses: Culinary. Petals eaten raw; used as tea. Stem and leaves edible as a salad green, vegetable. Dried stems are added to soups or stews. Leaves eaten raw, put in sandwiches or soups.

Nutritional Value: Leaves very high in beta-carotene, riboflavin, niacin and potassium. Low in carbohydrates and calories. Petals are very high in beta-carotene and niacin.

Collection and Storage: Gather petals only, leaving yellow centers.

Caution: Do not eat the yellow center, as it may cause indigestion. Collect clumps of new spring leaves for salads. Chew leaves or stems while hiking for a taste treat. Dries and freezes well.

Daisy

"She loves me, she loves me not!" We all enjoy and play this game, but think of how much more fun if you popped the succulent white petals in your mouth! Why not make a treat of the petals in salad or gelatin, or even in chocolate ice cream. I have put them in popsicles for kids.

Using the petals for food became a real survival tool when I spent hours picking blueberries or raspberries. They usually grow nearby, and when I found the petals edible, I could vary my diet from fruit to flowers. Then I discovered that daisy leaves are fantastic. From that moment on, Todd and I ate them while walking, playing, and working in the mountains of the Adirondacks. I can never look at a daisy and not remember the freedom of this period of my life!

Dandelion

Taraxacum officinale
Composite Family, *Asteraceae*

Other Names: Dent-de-lion, wild spinach, wild lettuce, red-seeded dandelion, wasanswak (Crow Indian).
History: Probably native to Europe, but widely distributed through the world. Used by many different cultures. Cultivated on the Mediterranean island of Minorca after locusts destroyed vegetation.
Habitat: Moist areas, lawns, roadsides, fields.
Characteristics: Perennial herb. Grows to height of 2 inches or more; clumped. Leaves saw-toothed; whorl from center. Yellow sectioned flowers mature to fluffy "pompoms" with seeds that blow in the wind.
Primary Uses: Culinary, medicinal. Early roots are eaten raw. Dried whole roots are pulverized or boiled. Leaves are eaten raw; steamed; boiled; sautéed; cooking liquid is drunk. Steeped for tea. Dried and ground for nutrient additive or flour. Flowers are eaten raw; sautéed; fried. Used for wine. Entire plant is steamed; boiled; sautéed; cooking liquid is drunk. Used in soups and stews.
Nutritional Value: High in protein, calcium, and vitamin A.
Medicinal Value: Use as tea for jaundice and other liver problems; a tonic, blood cleanser, purifier, stomatic, diuretic, laxative; for swellings, sores. Also effective for nervousness and hypochondria: an effective relaxant.

To reduce swelling and cleanse a wound, grind fresh leaves to a paste and spread over wound or fracture. Use dandelions as a bandage to bind wounds. Use as blood tonic by pulverizing either roots or leaves and mix 50-50 with whole wheat or rice flour. Make bread.

Cosmetic Value: Facials, wash for eczema; any external cosmetic skin treatments.

Collection and Storage: Gather flower heads in season. Yellow petals are very sweet. Twist the green base from the yellow flower petals and eat the petals raw. Roots can be dug year-round, even under the snow. Young leaves and stems are tastiest. Collect roots in early spring when most succulent and not bitter; dry and grind to coffee. Roots may be stored in strips (see page 30 for illustration on stripping roots). Store whole roots in paper bags in a warm, dry area.

Leaves
- 1/4 grocery-size bag = 8 cups leaves
- 8 cups raw leaves = 10 servings
- 8 cups steamed and frozen leaves = 4 servings
- 8 cups fresh leaves, dried on a screen = 1/2 cup flour

Flowers
- 1/4 grocery-size paper bag = 8 to 10 meals

Linda Says- Security—that is the only word I use for dandelion. Dandelion is always there, even frozen green under the snow, easily found, along edges of tree trunks or rocks. Ruth Spring taught me to core down the root, then peel the white succulent core for the first taste of a spring thaw. "As a tonic," she said, "the body will grab at it."

The former mayor of Vineland, New Jersey, started a large business growing and harvesting dandelions for the gourmet markets of Europe. The annual dandelion dinner he gave was nothing short of a gala feast.

Dock

Rumex crispus
Buckwheat Family, *Portulaceae*

Other Names: Curly dock, yellow dock.
History: Native to Europe, widespread in North America. Pima Indians use roots, stems, and leaves.
Habitat: Fields, roadsides, rich soil.
Characteristics: Perennial herb. Grows to height of 14 inches, may reach 24 inches. Elongated leaves curl; seeds turn dark brown when dry.
Primary Uses: Culinary, medicinal, cosmetic. Roots cooked as carrot. Leaves eaten raw; steamed; boiled; sautéed; cooking liquid is drunk. Stems eaten raw or cooked. Seeds ground for flour. Entire plant makes excellent yellow dye. Seeds are a decorative craft material, useful in flower arranging.
Nutritional Value: High in vitamin A, minerals, and protein.
Medicinal Value: Roots used for skin infections or itches, and for healing wounds and sores. Stems and roots are laxative and tonic. Pulp of leaves used as a poultice for bruises, burns, and swellings. Boiled and mashed, bound on boils and swellings.
Collection and Storage: You can spot dock easily in a field by its deep chocolate-colored tops. Several stands of dock are often in a field; the dark brown tops stand out in peaked clusters, especially above a winter's snow. Gather seeds when rust brown. Leaves are collected in early spring or when young. Roots are gathered in early spring for medicinal storage. Dries and freezes well.

The Department of Agriculture suggests collecting roots in the late summer and splitting while green for drying and grinding.

> **Linda Says-** I remember the summer Ruth Spring showed me dock. When I saw how many thick clusters of seeds there were on just one plant, I decided dock would be a main source of flour. I am constantly amazed at how many quarts of flour can be derived from one clump of dock. Because of the high vitamin A content of curly dock, this flour should be used sparingly only with another flour such as whole wheat. Curly dock seeds have 20,000 units of vitamin A per 1/2 cup. A few teaspoons in your recipe is adequate for vitamin content.

Evening Primrose

Oenothera biennis
Evening Primrose Family,
    Onagraceae

Other Names: Meadow rose.
History: Native to North Americas. Used by Native Americans.
Habitat: Gravelly dry soil, roadsides, waste areas.
Characteristics: Biennial night-flowering herb. Has pinwheel rosette of leaves that are toothless with whitish midrib. Basal leaves brown in spring, hugging the ground. Stalk shoots upward, with 4-petaled flowers on top of multiple branches. With 4 petals, the flowers are not unlike a rose, exceptionally sweet smelling and waxy. Hairy stem. Between stalk bottom and root underground is a strip of red or pink to scarlet indicative of primrose. Single root with many hair-like extensions.
Primary Uses: Culinary, medicinal, cosmetic. Young leaves are eaten as a vegetable or salads. Dried and frozen for storage. Buds edible before flowering. Seeds edible. Roots look and taste like parsnip (peppery flavored), may be eaten raw or cooked.

Dried for long-term storage.
Medicinal Value: Evening primrose oil used for soothing the central nervous system as well as relieving stomach and gastrointestinal stress. See below.
Cosmetic Value: Creams, skin preparations.
Collection and Storage: Pick young leaves in early spring. Pinch flowers off stem. Dig roots and scrub. Roots can be dug in winter.

> **Linda Says-** Rumor has it that oil from this delectable wildflower is one of the most expensive, as well as most beneficial for human beings. Its composition close to the constituents of mother's milk, evening primrose offers many medicinal remedies.
>
> I began picking and eating the flowers in quantity. I use olive oil to make evening primrose oil, just press the flowers into a small jar and add boiling olive oil. (Be careful to avoid splatters.) Press the flowers down with a spoon and pour in oil until flowers are saturated and air has bubbled up. Cap tightly and shake every day or so for a week. Strain the oil and refrigerate. It's a potent tonic when added to your salads.

Filarie

Erodium circutarium
Cranesbill Family, *Geraniaceae*

Other Names: filaree, alfilaria, redstem, stork's bill, pin clover, alfilaree.
History: Native to Europe and probably introduced to North America by the Spaniards; Native American use.

Habitat: Dry areas, desert areas with a little moisture, particularly clay soils.

Characteristics: Annual or biennial, reproducing via seeds at the base of stork's bill-shaped shoots. Flowers on long stalks shoot up from fern-like leaves lying in thick mats. Minute hair on stalks and stems. Stalks have purple flowers, then seed buds. Seed has a corkscrew tail, driving it into the soil for propagation.

Primary Uses: Culinary. Fern-like leaves are eaten raw like celery. Used in soups, stems, sandwiches. May be steamed and frozen for summer vegetable. Roots are like celery, delicious raw or in soups; steamed or pickled. Both roots and leaves are dried and ground to a flour.

Nutritional Value: Leaves are very high in calcium, potassium and beta carotene, niacin and vitamin C.

Collection and Storage: In the west I used a gloved hand, taking the most succulent leaves only up into your hand, pulling them upwards, and with a twist of your wrist, twist the leaves off the root. This leaves the root and flower stems.

Caution: Do not ingest flowers or stork's bill fruits, as they are known to be toxic. It is generally safe to ingest a large amount of filarie leaves if the vein in the center of the leaf is not pink or red. Filarie tends to pick up nitrates from the desert soil, turning the vein of the leaf as well as the stem of the flower red. Look carefully for healthy plants, green and luscious before eating.

Linda Says- A winter's filarie sandwich has become a ritual. The greens appear after the rains on the Arizona desert. I clip them with scissors, wash them, and make a mild, nutritious raw vegetable served with a dip. Pickled in apple cider vinegar, they make a delicious salad ingredient.

Fireweed

Epilobium angustifolium
Evening-Primrose Family,
    Onagraceae

Other Names: Great willow herb, deerthorn, wild asparagus, wickup ocacadii (Potawatomi).

History: Native to North America. Staple food in Northern Asia and Iceland. First plant to re-seed itself after all wars of wasteland or defoliation. Grew on sides of Mt. St. Helen 3 weeks after volcano covered area in potash. Grows back first after fires. The Potawatomi Indians used its root as a soap by lathering it in water.

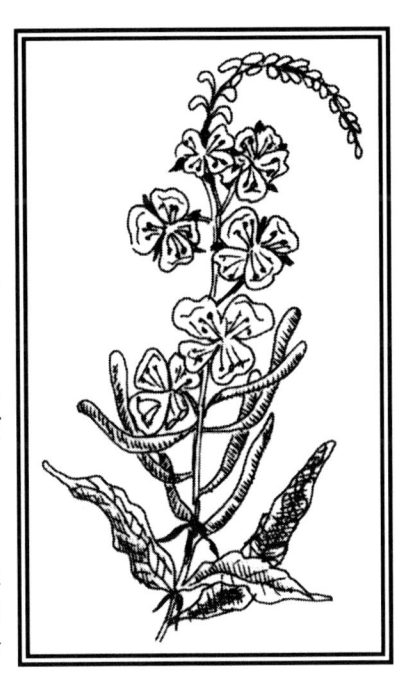

Habitat: Areas where there has been fire, waste areas, areas of regrowth, roadsides, fields.

Characteristics: Perennial herb with unbranched tall spikes up to 2-6 feet and willowlike. Spikes not unlike upside down asparagus. Stems hollow with white pith.

Primary Uses: Culinary. Pith is eaten raw or boiled; used in soups and stews; steeped for tea. Buds are eaten like asparagus; tips picked before they flower are eaten raw or used in soups and stews. Flowers may be eaten, but buds are more palatable. Young leaves eaten raw or prepared as spinach; dried for tea. The fluff makes excellent winter tinder; used as cotton.

Nutritional Value: Exceptionally high in beta carotene and potassium as well as most minerals.

Collection and Storage: Perennial herb. Reaches height of 2 feet or more. Spike of magenta or pink 4-petaled flowers; elongated top of buds.

Linda Says- The best way to see the old fireweed beds is when it is snowy. The plants' spiral top, tall and fuzzy, swirls with cotton fuzz. In season, fireweed shows a magenta top with flowers. Above the flowers are asparagus-like tips that I harvest as a delicious vegetable. They taste best fresh.

Goldenrod

Solidago odora
Sunflower Family, *Compositae*

Other Names: Golden elixir, sweet goldenrod, asawush (Ojibive Indians). Poco moonshine is yet another species, *augustifolum*.
History: Native of North America. Used by pioneers. Ojibive Indians used it in tea as a carminative, antispasmodic, and intestinal astringent (Smith 1932).
Habitat: Dry, open fields, woods, road banks, pine barrens; acidic soil.
Characteristics: Perennial herb. Grows to height of 24 inches or more. Hundreds of tiny fluffy yellow flowers on top stalk. Flowers smell like anise.
Primary Uses: Culinary, medicinal, cosmetics.
Roots eaten raw; dried and pulverized for tonic powder. Stems dried for tonic nibble. Leaves picked and eaten fresh. Dried for tonic tea. Stems, buds, flowers, and seeds eaten or used as tea.
Medicinal Value: Energizer and tonic. As a tea, for diarrhea, bladder tone, colic. Mashed leaves for a poultice, raw flowers

chewed for sore throat. Leaves and roots are tonic for colds and upper respiratory problems. Powdered leaves are sprinkled in sores and wounds because goldenrod is an astringent. Helps stop internal hemorrhaging as a strong tea.

<u>Cosmetic Value</u>: Shampoo, astringent with diaphoretic qualities; aromatic and stimulating.

<u>Collection and Storage</u>: Gather the entire plant, bundle it, and hang it upside down until dry in a warm, airy place out of the sunlight. The bundles are dry when the stems snap easily. Place a clean sheet under the bundles. Hand strip and gather up the bounty. To store a winter's supply of tea, gather the stripped and dried goldenrod parts in a pillowcase or other cloth bag, then wring to crush the leaves. A handful of flowers and leaves will brew 6 to 8 cups of tea.

<u>Caution</u>: Do not use as bulk food. Use sparingly for medicinal purposes.

Linda Says- So easy to gather and bundle, the golden elixir grows thick and bountiful. Although its tonic qualities are present any time, the goodness is strongest just before the pollen forms in the flower head. If you choose to collect at this time, watch out for the bees! They love the flowers and frequently hide in the caps or rest in them.

For a perfect yellow tea, snip off the blossoms. Dry separately on a screen and store in glass bottles.

Grape

Vitis arizonica (Western)
Vitis labrusca (Eastern)
Vitis rotundifolia (South)
Vitis Family, *Vitacere*

Other Names: Wild grape.
History: Native to North America.
Habitat: Thickets, edges of woods.
Characteristics: Deciduous vine. Forked tendrils at end and joints of high-climbing vines. Branches have a brown pith. Leaves tiny to large, coarse, saw-toothed, lobed, heart shaped. Fruit round, fleshy. One to 4 pear-shaped seeds in several colors from purple or black to amber.
Primary Uses: Culinary. Fruit cooked to make jellies, jams, juice, syrups, and flavorings. Leaves are used as a wrapper for main dishes. Leaves may be eaten raw, stir-fried, or steamed. Tendrils are pickled.
Nutritional Value: Fruit is high in vitamin C.
Collection and Storage: Pick tendrils all year; clip with scissors. Grapes in season may be gathered by hand. Young leaves are picked at any time. Freeze leaves for stuffing; laying between sheets of wax paper, folding paper after each layer.

Linda Says- I wandered the hills of Syracuse, New York, with a famous gourmet cook looking for grape tendrils. There they were, curling down from yards of grape vines. I rolled and sniffed them first for their characteristic grape smell, then used scissors to clip hundreds of tendrils into a bowl. Later, my friend and I stir-fried the tendrils into delicacies, as well as froze them for use later in the year. The tendrils make me pucker.

Lamb's Quarters

Chenopodium album
Goosefoot Family, *Chenopodiaceae*

Other Names: Pigweed, goosefoot, wild spinach, chiciquelite, hwahai (Pima Indian). (Toakush Crow Indian).

History: Used in most countries and most civilizations. Recent archaeological findings in Denmark show it was an early food there. Now cultivated by Navaho and Mexicans. The Zapotec Indians use the flour as a staple food.

Habitat: Native to North America, Europe, Asia.

Characteristics: Annual herb. Grows to height of 18 inches or more. Leaf shaped like goose's foot, dark green with whitish underlay. New leaves have white or lavender-tinged powder near center whorl. Leaf beads with water when wet. Green seed-like flower clusters.

Primary Uses: Culinary, medicinal. Young shoots are boiled, cooking water is drunk. Leaves are spinach-like; eaten raw; steamed; boiled; sautéed; cooking liquid is drunk. Dried and ground for flour. Flower and seeds are ground for mush or cereal; sprouted, dried and ground for flour. Seed is excellent food for wild birds and canaries. Roots can be used as a soap substitute; wet hands and rub cleaned root. Young stems eaten raw or cooked.

Nutritional Value: High in nutrition, vitamins, calcium.

Medicinal Value: All parts used as a poultice for swelling, rheumatism, arthritis. Chewed raw for toothaches. Also, gelatin capsules filled with lamb's quarters are a potent vitamin.

<u>Collection and Storage</u>: Harvest large amounts with a long knife or small amounts with scissors and bag. Hang large bunches upside down to dry. When crispy dry store in glass for longevity or steam and freeze.
- 1 large bunch (stems bunched together to about 1 inch thickness) = 2 cups flour when dried and stripped.
- 1 large screen of leaves and seeds = 2 cups flour
- 1 large bunch chopped and steamed = approx. 12 to 14 sandwich sized plastic bags.

<u>Caution</u>: The lamb's quarters' leaf resembles poisonous, malodorous look-alike, nettleleaf goosefoot, which has a tough cardboard-like leaf. Crush and smell leaf. The rank odor will tell you immediately if you have this look-alike. Identify carefully. See page 294.

<u>Comments</u>: Chenopodium ambrosiodes; also known as Jerusalem oak, wormseed or Mexican goosefoot, is a type of lamb's quarters found in Arizona and other Western states. A sprig added to beans renders them anti-flatulent.

Lamb's Quarters

During World War II, the US government gave out information that garden weeds were good food. The top weeds included amaranth and lamb's quarters. Not only are these plants nutritious, but they are not bothered by cutworms or aphids.

Lamb's quarters grows phenomenally fast. In fact, just one plant yields all I need for a summer's meals for me! To gather, I pinch the tops and tips of each branch. I wait two or three days, then repeat; for each top taken, two or three will grow in its place. So just a few plants can yield an entire fall and winter supply.

Dried leaves make a delicious and nutritious flour. Mix it with a bit of water to make a tortilla, or combine with wheat flour to make an enriched pita bread or fry bread.

To make the flour, I pound the dried leaves with my favorite rock, but you can feed the leaves through a hand grinder for the same results. You can also place the crispy dried leaves in a pillow case and wring.

Malva

Malva Neglecta
Mallow family, *Malvaceae*

Other Names:
Cheeseweed, button weed, cheeses, mallow.
History: Introduced from Europe; sold in the marketplaces of China and Japan.
Habitat: Fields, roadsides, waste areas.
Characteristics: Perennial or biennial herb. A stout, bushy plant branched and spreading from its base. Reaches height from 2 feet to 4 feet.
Primary Uses: Culinary, medicinal. Leaves used in soups, main dishes, stir-fried. Stems are vegetable stir-fried in garlic and olive oil. Flowers, buds and stems, and cheeses (seed cases) are all edible raw, steamed, sautéed, or dried and stored.
Nutritional Value: Very high in iron and calcium.
Medicinal Value: A mucilagenic, increasing the flow of saliva. Chewed for sore throats, upper respiratory problems.
Collection and Storage: Clip whole stems, pick off buds and leaves. Cut stems into sections for stir-fry or freezing. Cut leaves as large as you can for rolling around tofu, fried beans, potatoes, rice. Collect in bunches for abundance. Do not pick malva with red stems. Red stems may depict nitrate from the soil; pick fresh young plants when possible. Mince malva leaves and marinate in oil and vinegar before cooking.

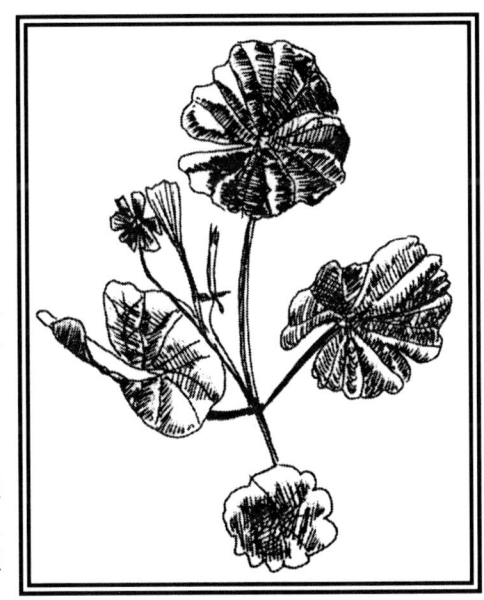

Add 1 cup salsa to ice cube trays, a spoonful of fried beans, roll tightly in malva leaves and freeze. Easy to defrost and bake at a later date. (See recipe for Malva Rice Roll-Ups.)

Stuffed Malva Leaves

Linda Says- Malva has cheese-wheel seeds. I love to gather them and dry them for potato salads, stir-fries, or eat them as they are.

Seeds are dried to use as "nuts" in winter. Large leaves used for wrapping foods. Stems are used for delicate sautéed vegetable, and younger leaves for salads and sandwiches. Once dried, all parts of malva can be ground for flour! I cook with malva flour a lot. As a general rule, I use 1 cup malva flour to 3 cups of wheat or other gluten flour.

Maple

Aceraceae Family

Other Names: Sugar, silver, soft, rock and red maple.

History: Many species native to United States. Native Americans were early users of the tree's products. Chippewa Indians pressed the seeds into cakes calling them maple cakes.

Habitat: All environments except lower Rocky Mountains and the Plains states; areas with rich, well drained soil.

Characteristics: Deciduous tree. Most leaves have 5 pointed lobes, are irregular toothed, and are opposite on branches. Leaves turn golden yellow to scarlet in early fall. Seeds have brown wings and occur in pairs. All have sweet sap to a varying degree.

Primary Uses: Culinary. Sap used for syrup, candy, flavorings. Seeds eaten raw or pressed into cakes. Twigs and young leaves used for tea and emergency food.

Nutritional Value: Leaves high in minerals, beta carotene and vegetable protein.

Collection and Storage: Cut the seed off the wing with scissors, dry on a screen, and put through the grinder. Twigs, inner bark, and sap can be collected and used in winter. Collections of twigs is simple because they snap off easily by hand. Syrup collection is through a tap drilled into tree using a collector pail. (Author used plastic milk bottles from a nearby dump.) Please refer to other references for details of syrup collection.

Maples in Winter

I was relaxing in front of a fire in the crispness of early morning when Crack! A sound like an explosion came from behind me in the woods. I scanned the trees and saw that a maple tree had "exploded." The explosion caused a big crack in the tree about three feet high. When a winter wind stirs the frozen trees, they sometimes appear to burst vertically. When it was 40 degrees below zero at night, I lay awake and listened to trees explode. That's a true wilderness thermometer!

Meadowsweet

Filipendula ulmaria
Rose Family, *Rosaceae*

Other Names: Queen-of-the-meadow, meadow sweet.
History: Native to North America, used by native Americans.
Habitat: Meadows, old pastures from Nova Scotia and New Brunswick to North Carolina and west to Minnesota and Arkansas.
Characteristics: Perennial herb. Bush with symmetrical branches, 30 inches long or more. Flower spires, shaped like church steeples, are white and fluffy in spring. The steeples turn seedy and chocolate brown in the winter, giving it a characteristic look against the snow. Twigs have hundreds of tiny nodes between the branches and are characteristic of meadowsweet, not other hardhacks.
Primary Uses: Culinary. Twigs chewed or steeped as tea. Dried and stored for winter. Flowers are dried as a sugar substitute; sweet enough for teas & cereals, but flower must be dried. Used in wines. Seeds are also edible. Leaves used for tea.
Nutritional Value: High in vitamins and minerals. Also high in beta carotene, potassium, and niacin.
Medicinal Value: Mineral tonic. Flu and Rheumatic pains. (contains salicylates.) Twigs, inner bark and sap can be used in winter.
Collection and Storage: Collect twigs by breaking off when young and store in paper bag. Collect flowers by pinching off easily with fingers. Dry on screens.

Linda Says - Shearing meadowsweet flowers with scissors is by far the quickest way to harvest them. Pick with both hands, breaking brittle twigs. Swing a handful from your right hand up under your left armpit. Hold tight and continue until bunch is cumbersome, then tie and set aside. Gather bunches and tie together.

Milk Thistle

Silybum marianum
Composite Family,
Compositae

Other Names: Horse thistle, marian, holy thistle.
History: Native to the Mediterranean, also hot regions of North America, including Southern Arizona. Cultivated and dried in Europe. Used by Native Americans.
Habitat: In west, commonly found in ditches; waste places.
Characteristics: Annual herb. Grows 3 to 6 feet tall. White milk splotches on prickly leaves, named "Mother Mary's milk." Large prickly, fuzzy thistle flower with long prickers from central stem. Vigorous grower; 1 plant will quickly spread to a bed 3 feet around.
Primary Uses: Culinary, medicinal. Young leaves parboiled to remove prickers, for soup or stir-fry. Cooking liquid is drunk. Leaves dried and ground for extract.
Nutritional Value: High in vitamins and minerals. High in phosphorous, vitamin K, and thiamin and niacin.
Medicinal Value: For treatment of liver disease. A valuable source of the chemical silymarin. Mineral tonic, flu and rheumatic pains (contains salycylates).
Collection and Storage: Use a long blade or sharp machete to cut off prickly leaves. Extract young leaves from center of plant with long knife and tongs. Plant tends to pull up nitrates from the soil, so harvest young leaves. I dry the leaves by placing them face down on a screen. You can handle prickly leaf safely by pinching the back vein of the leaf.

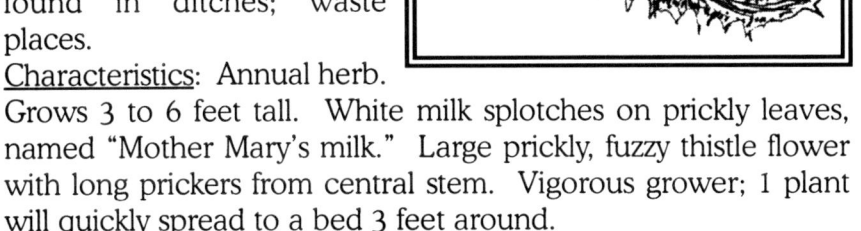

Linda Says- The gigantic prickles on the milk thistle once scared me. I was aware of the multiple benefits of these ominous leaves, but how to gather them? Gloves did not help. The best tools are a long-handled tongs and a long kitchen knife. Slice the leaf off with the knife and pick it up with your tongs and bag it. Washing them becomes the next challenge, and cooking to a safe texture is yet another. The leaves fairly "melt" in cooking; even the barbs become tender if the leaf is simmered long enough. I cut off the barbs with scissors, then simmer a few minutes and drink the broth.

Milk Thistle

Milkweed

Asclepias species
Milkweed Family,
    Aslepiadaceae

Other Names: Silkweed, butterfly weed, cotton tree, pleurisy root.
History: Native perennial with many species. Known since ancient times, used during World War II in rubber experiments, fluff considered for life preserver stuffing.
Habitat: Fields, roadsides, gardens.
Characteristics: Perennial herb. Single stem 2 to 5 feet high. Leaves opposite in pairs from 4 to 9 inches long. Heavily veined, smooth leaf with white milky sap throughout plant. Flowers are round clusters of pleasantly fragrant single stars at end of dozens of branches, making a round ball of blossoms. Flowers give rise to a green bristly pod, in which fluff-bearing seeds form. White sap exudes from all parts, including roots.
Primary Uses: Culinary and medicinal. Flowers and seeds are edible. New sprouts are used as vegetable, cooked. Young leaves and buds are cooked as vegetable. Stalks are peeled for strong fibers; used for fishing line, especially if braided. Bark is braided into ropes of varying thickness. Pods picked young are used in soups and stews.
Nutritional Value: Shoots high in niacin, potassium and vegetable protein.
Medicinal Value: Milky sap used in wart removal.
Collection and Storage: Break off buds, young leaves, and flowers by hand. Use a paper bag. The milk tends to turn the plant sour if kept in plastic bags. Strip stem fibers and braid for emergency fishing line or rope, as well as shoelaces (in the wild).

As a vegetable; young leaves and buds can be dried or frozen. Sprouts can be frozen. Buds should be simmered twice for 3 to 4 minutes, changing water in between. Roll any wide leaf with goodies and bake in salsa for a wild Mexican delight.

Wild Mexican delight!

Milkweed Abundance

Gathering milkweed is by far the easiest way to find food for my family. The abundant pebbly tops beckon to me. I move from every fourth plant or so with ease. I fill a large bag and then go to the Cedar River log bridge to wash them. The time spent swishing the milkweed buds in the cool water are some of the most precious moments of homesteading.

Mint

Mentha species
Labiatae family

Other Names: Spearmint, peppermint, horehound, giant hyssop, wild bergamont, bee balm, oswego tea.

History: Mostly native to North America and widely distributed throughout the world. Used since ancient times in scents, cosmetics, flavorings, food and medicines.

Habitat: Found in most areas from swamps to fields, gardens to lawns.

Characteristics: Perennial herb. All mints are aromatic. Most mints have square stems; leaves are opposite with usual unevenly saw-toothed edge. Prominent veins; leaves may be smooth or coarse, sharply pointed in most species. Characteristic fuzzy leaves in wild varieties, especially swamp mints. Plants range from 5 inches to 2 feet high.

Primary Uses: Culinary, medicinal, cosmetic. Leaves, flowers, stems and roots eaten raw in salads, candies; leaves steeped for tea. Flowers dried for teas. Whole plant ground for aromatic tea.

Nutritional Value: High in vitamin C and iron.

Medicinal Value: Well known as a stomachic, or stomach soother. Also used for insomnia, stomach ache, fevers and colds.
Cosmetic Value: Aromatic stimulant, astringent, restorative; used for scents, in soups and potpourris.
Collection and Storage: Cut stems where desired, as plant rejuvenates quickly. I used to pick and wind one plant into a bunch, then string bunches together for ease of harvesting as well as drying. Also, freeze flowers in ice for pretty edible cubes. Mints transplant easily and have characteristic prolific growth in an herb garden.

Linda Says- The camp always smelled good when the mint was drying for winter. Our family eats the leaves often. Since a leaf holds a whopping amount of iron, it is understandable why I eat at least 1/2 cupful every time I go out to pick it.

Mint Aromas

Lightning flashing all around me, I lay prostrate in a large patch of mint. The air was blue-black. Tired and clutching my bundle of freshly picked plants, I almost welcomed the respite, though I was a bit nervous.

Mint picking is always done around swamps, streams, and riverbeds, and this day I was able to gather a six-week supply before a typical Adirondack thunderstorm kept me waiting for 30 minutes.

Mullein

Verbascum thapsus
Snapdragon Family,
Scrophulariaceae

Other Names: Candlewick, flannel leaf, Aaron's rod, feltwort, hare's beard; velvet plant, velvet dock, blanket leaf.

History: Native to Europe, naturalized throughout pasture land of United States. Plant also known in Roman and Greek history; Quakers, Native Americans.

Habitat: Waste wet areas as well as roadsides, gravel-filled fields.

Characteristics: Biennial herb. Tall spikes with yellow flowers, velvety leaves whirl around spike. Grows up to 10 feet tall. Produces basal leaves in first year, then spike the second year. Leaves furry or velvety, ranging from 3 inches to 2 feet long. Flowers are succulent and sweet smelling, attracting bees.

Primary Uses: Culinary, medicinal, cosmetic. Root used for medicinal tea, leaves and flowers dried for tea and medicine. A normal root, up to 1 foot long, may be reused dozens of times. I put large leaves in my sneakers when walking long distances; they help soothe tired feet, and prevent blisters. Native Americans put leaves on cradle boards to heal elbows and shoulder blades. Antiseptic, mullein is invaluable for inflammation.

Medicinal Value: All parts of mullein, from single tap root to flowers have an antihistamine quality among other qualities. (See chart page 53 for comparing mullein to aloe vera.)

Cosmetic Value: Astringent and emollient properties, decoction of flowers for hair rinses to lighten hair color; flowers steeped in olive oil used as application for sores or massaging oils.

Collection and Storage: Second-year leaves may be harvested from the spike individually, flowers harvested individually, and root harvested at any time, any season. I harvest single leaves, roll them up, and dry them on a screen. These rolls may be kept in glass containers and taken along when hiking.

Floated in olive oil, mullein wicks will burn cleanly & clearly for 3 hours or more or until the olive oil is used up. Pour a tablespoon of olive oil on the top of a glass of water. (You must float the mullein by crossing 2 toothpicks, poking the ends of the toothpicks in a tiny piece of foam or buoyant material.) Place the wick in the center, letting a piece touch the oil you have poured on the top of the water.

Figure 48 Rolling and Drying Mullein leaves.

Linda Says- I used mullein for lamp wicks when snowed in. A small piece of the flannel leaf cut to size, dried, and soaked in oil will burn for quite a while.

Mullein

I feel as if the 6-foot high plant were a person. I address the mullein and ask to pick its flannel leaves and flowers!

One summer, my Girl Scout Troop was horseback riding on a field trip in Indian Lake, New York. One of the horses moved too close to the one in front and the forward horse kicked, clipping the child on the ankle. It was obvious she had a bad injury. Removing her shoe, I shouted to get mullein, quick! The scouts came running with fresh mullein. "Roll the leaves and crush them quickly with a rock, girls." Poultices were made and applied on the swollen ankle. The mullein stalks became splints and the leaves became bandages until we could get the small scout to a hospital. Soon thereafter a doctor called to ask "What was that plant around the scout's ankle? She had a chip fracture but no swelling or pain!" Mullein, sir!

Mustard

Brassica species
Mustard Family, *Brassicerceae*

Other Names: Charlock mustard, field mustard, yellow mustard.
History: Native to Europe, pasture land in North America. Used medicinally in Greek, Anglo-Saxon history. A pot herb in English history.
Habitat: Fields, roadsides in cooler regions.
Characteristics: Annual herb. Grows to height of 10 feet or more. Leaves rounded, with extra protuberances of tiny leaves below main part of leaf. Crushed leaf yields herby pungent smell. Four-petaled yellow flower; petals in form of cross. Black seeds in pod.
Primary Uses: Culinary. Stems used raw as pungent spice. Leaves all eaten raw, used as pungent spice; steamed, boiled in soups, stews. Flowers are eaten raw; steamed. Seeds dried and used as spice; ground for mustard.
Nutritional Value: High in beta-carotene, minerals, vegetable protein.
Medical Value: Dried leaves and flowers used as a poultice, or mustard plaster for respiratory distress.
Collection and Storage: Collect young leaves by snipping off stalk with fingers. Collect pods separately in season when seeds are mature, pods almost dry.

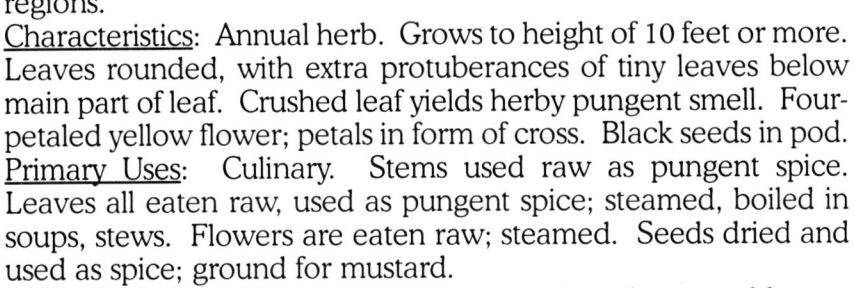

Linda Says- When eating mustard greens boil in water for 15-20 minutes. Mustard is a very pungent vegetable.

Mustard

As soon as the temperature rose to between 32 and 40 degrees, I would make a beeline to the back of Alice and Walt Sherman's barn. The mustard plants were always green, growing beneath the snow against the warm barn boards. There are no words to describe the taste of the first mustard leaves of the season. My body fairly roars with energy for days to follow!

Nettles

Urtica species
Urticaceae Family

Other Names: Stinging, slender, great, dwarf, nettle.
History: Native to Europe, widely naturalized in North America. Used worldwide.
Habitat: Fields, fertile soil from Alaska throughout continental US.
Characteristics: Perennial herb. Leafy plant. Leaves opposite in pairs, coarsely veined, oblong tapered to tip with rough, sharp saw-toothed edges. Entire plant fuzzy with tiny stinging bristles or minute hairs. Flowers are tiny green seed-like clusters between stalk and branches.
Primary Uses: Culinary, medicinal, cosmetic. Leaves and flowers boiled for soups and stews, cooked in stir-fries. Seeds are used in herbal teas. Excellent tea!
Nutritional Value: High in protein, iron and vitamin C.
Medicinal Value: Tea well known as a nutritious restorative tonic.
Cosmetic Value: Commercial uses: nettle shampoo, herb wraps.
Collection and Storage: Use leather gloves when collecting nettles. When harvesting for vegetable, hold a bag or bowl under separate leaves and clip off with a pair of scissors. A few plants sustain a whole family, since several nettle leaves grow where pruned.

Clip seeds separately if desired, plant will continue to produce until mature. Place nettles in water to simmer away hair and leave a nutritious main dish.

Linda says- I can still feel that prickly burn on my leg from each tiny hair as I approach the tall, stately plants. I learned my lesson. Blessing the "armored" leaves, I soon practiced safe collection. My young scout troop once gathered a bunch of nettles and presented a "bouquet" to Utica's city scouts; I'm sure their lesson became the same painful memory.

Phragmities

Phragmities communis
Grass Family, *Graminae*

Other Names: Reed, giant reed.
History: Native to Europe and North America. An ancient food, craft material, fiber. The Gosiute Indians gathered a sweet secretion formed on the leaves by aphids and used it as sugar (Chamberlain 1911). The Panamint Indians dried the entire reed, ground it, and sifted out the flour (Colville 1892). This moist and sticky substance was then set near a fire until it swelled and browned, when it would be eaten like taffy. The best time to gather the culms (stalks) was spring.
Habitat: Grows throughout the United States except inland areas of South Atlantic and south-central states; also Europe, Asia. Likes marshes, swamps, wet areas, roadsides.
Characteristics: Tall, reed-like stalks with feathery seed tips. Up to 6-8 feet tall.
Primary Uses: Culinary. Stalks are dried and ground to flour. Gummy substance in the stalk can be used as gum or in drinks. Stems woven for mats, bedding, thatching, grass huts. Excellent tinder. Fibers in stalk made into ropes by twisting together. Seed heads are made into a gruel, cereal, or flour. Used in cellulose, arrow shafts. If seeds are not present, chaff is ground into flour. Roots or rhizome (underground stems) are gathered and ground into flour. Eaten raw, roasted, or boiled into a sweet, gummy confection not unlike candy or marshmallow.
Collection and Storage: Study the cycle of head (seed) maturation. There is a 2-week period when seed head is full and ripe. Past that stage the seeds drop and blow away. Clip heads with scissors in the middle of summer, before seeds mature and fall out of seed head.
Caution: The water in which phragmities grows is often polluted, so test the water before eating the plant.

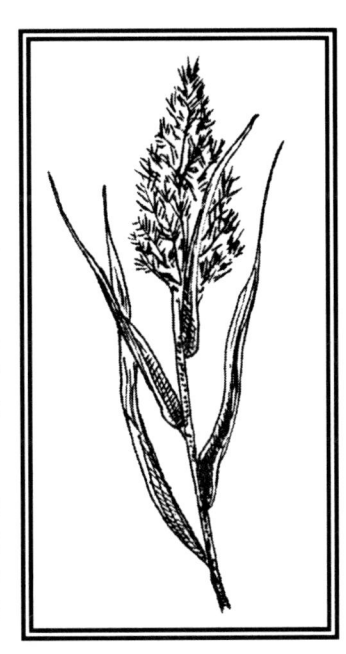

Linda says- Imagine my surprise when told that the waving reed was edible and could be ground to flour! It wasn't long before I began to experiment. The thought that such a tremendous proportion of the world was home to this plant, but few people were baking with it became another reason for this book. (The water in which phragmities grows is often polluted, so test the water before eating the plant.)

Pine

Pinus species
Pine Family, *Pinaceae*

Other Names: Too numerous to mention, but including white pine in East and pinyon in West.
History: Native to most parts of the world. Used by Native Americans.
Habitat: Woods, forests, landscaped area; soil varies from dry to moist.
Characteristics: Evergreen tree. Cone-bearing. Needles vary in length, usually long and slender occurring along the twigs. Bundles of 2 to 5 needles common. (White pine with 5 needles shown in picture.)
Primary Uses: Culinary, medicinal, cosmetic. Needles eaten raw or cooked as vegetable; tea, flour. Catkins eaten as candy; raw, dried for tea, salads. Cone and nuts (seeds) are eaten as snack; ingredient in baked goods. Bark is emergency food, roots are eaten raw, steeped for tea. Sap has medicinal application for small cuts, scrapes, blisters.
Nutritional Value: High in Vitamin C.
Medicinal Value: Colds, flu; as a tea, mixed with honey.
Cosmetic Value: Deodorant, air freshener; soaps, lotions, cleansing oils.
Collection and Storage: Gather needles, twigs, and bark. Using a knife, peel a section of bark lengthwise from a young branch; this may be chewed for a long time. (See "Peeling bark in strips" on page 31.) To harvest needles, let branches dry thoroughly

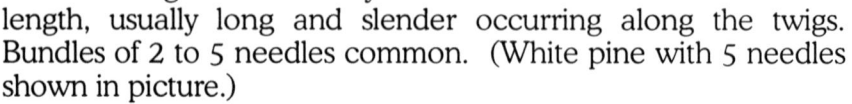

until needles fall. Gather needles in a sheet, then break up twigs into small sections. Place needles in glass jars. Preheat oven to 300 degrees, then turn off! Place jar with needles in oven and leave until the oven is cool. Cap immediately and store.

To do twigs, place them in a glass jar, leaving the cap off. Put in a turned off 300° F. oven, leaving until cool. Put in a sunny area for further drying, if needed (until they snap easily). Cap for storage.

Caution: Needles are volatile and catch fire easily.

Linda says- Many was the time I ploughed through snowdrifts looking for meadowsweet, maple and beech twigs, and succulent fresh pine needles. Pine provides the energy to go on! It is the taste and smell of the forest. Save every bit of your Christmas tree for pine tea if it has not been sprayed or dyed a darker green.

Plantain

Plantago major (common plantain)
P. Juncoides (Long-Leafed Plantain)
Plantain Family, *Plantaginaceae*

Other Names: Turnip leaf, cart-tract plant, cuckoo's bread, English plantain, goosetongue, Indian wheat, pale plantain, ribwort, rippleseed plantain, seaside plantain, seashore plantain, snake weed, soldiers herb, White man's footprint.

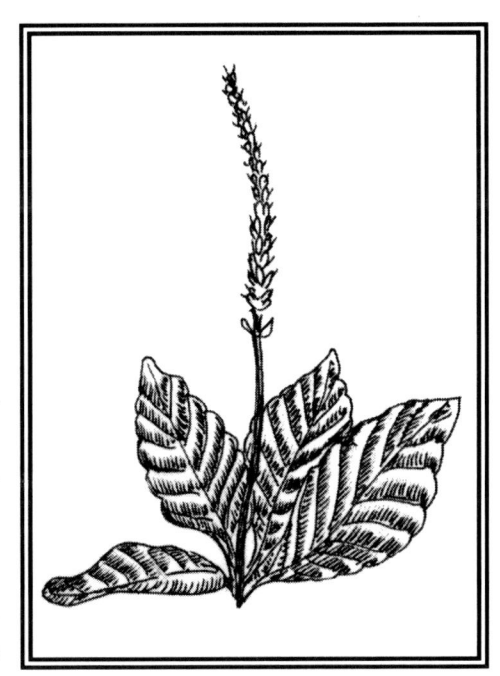

Common Plantain

Long-leaf Plantain.

History: Many types native to Europe and naturalized in North America. Used as vegetable in France; cultivated in England. Some varieties native to Asia.

Habitat: Lawns, roadsides, fields, waste areas.

Characteristics: Biennial or annual herb. Turnip-like flat roundish leaves, with prominent parallel veins. Center stalk lined with seeds. Considered a pot herb in Europe, this plant has many species. Common plantain has spade-like flat leaves, whorling on the ground from a center point, with tall stalk in center with many seeds around stalk up to tip. Long-leafed plantain has long narrow leaves with heavy vein and similar texture, taste, and smell. Stalk has seeds at only top inch or so.

Primary Uses: Culinary, cosmetic. Young leaves are eaten raw or cooked. Stalks are dried for vegetable chewies. Seeds are eaten as sprouts. Entire plant is edible.

Collection and Storage: Turnip-like leaves are easily twisted from the ground by hand. The stalks are clipped and dried in a day or so. Use scissors. Leave stalks until the seeds are easily removed. Save stalks for a crispy stick food.

Linda says- The plantain leaf I eat is also a fantastic underarm deodorant! Prepare the leaves by placing 3 leaves together, then crush with a rock or hammer until the juices flow. Remove the inside leaf and wipe juice on underarm. Free supply abounds under my feet. Plantain has surrounded all the camps I have lived in throughout my homestead years.

Prickly Pear

Opuntia megacantha and other species
Cactus Family, *Cactaceae*

Other Names: Devil's tongue, beavertail, Indian fig, tuna.
History: Native to North and South America. Used by Native Americans.
Habitat: Hot, dry, well drained soil.
Characteristics: Fleshy, low-spreading cactus. Has flat stems and pads with long needles. Thorny fruits are preceded by waxy flowers, yellow or red. Fruits ripe for harvesting are yellow-green to purplish black.
Primary Uses: Culinary, cosmetic. Pads are cut up for food; strips parboiled and french-fried; simmered for hair tonic and softener. Flowers eaten raw, cooked to candy, jelly; also frozen in cubes for drinks. Buds cooked for jellies, jams, syrups, deserts, candies; juices also made. Water from fruit and young pads may be used as a liquid in an emergency. Seeds may be dried and ground to flour; used as a soup thickener.
Nutritional Value: High in carbohydrates and calories.
Medicinal Value: Pads split open for soothing gel for burns, wounds, sunburn.
Cosmetic Value: Hair tonic, softener.
Collection and Storage: Handle pads only with leather gloves and tongs. To harvest, use a long stick, handle, or shovel to knock off each pad. With a deliberate touch-and-push movement, the pads break off and fall to the ground with ease. When you pick them up, slide a board underneath or use tongs, and put them in a container. One pad is a large filet, or 2 servings. It can be cut up and frozen in cubes for drinks.

One hair rinse pad is good for several shampoos. Fruit and seed pulp also harvested with tongs. Flowers are easily removed by hand, petal at a time. Petals eaten raw, cooked to candy, jelly.

To prepare pads as poultice, hold gingerly with thumbs and forefinger. Using point of a knife, poke first, then slice carefully in half. Spread out and scrape off watery meat for use as a poultice.

<u>Caution</u>: Prickly pear is a protected plant. OK to harvest pads, flowers, buds as a survival food on the desert. Do not remove whole plants.

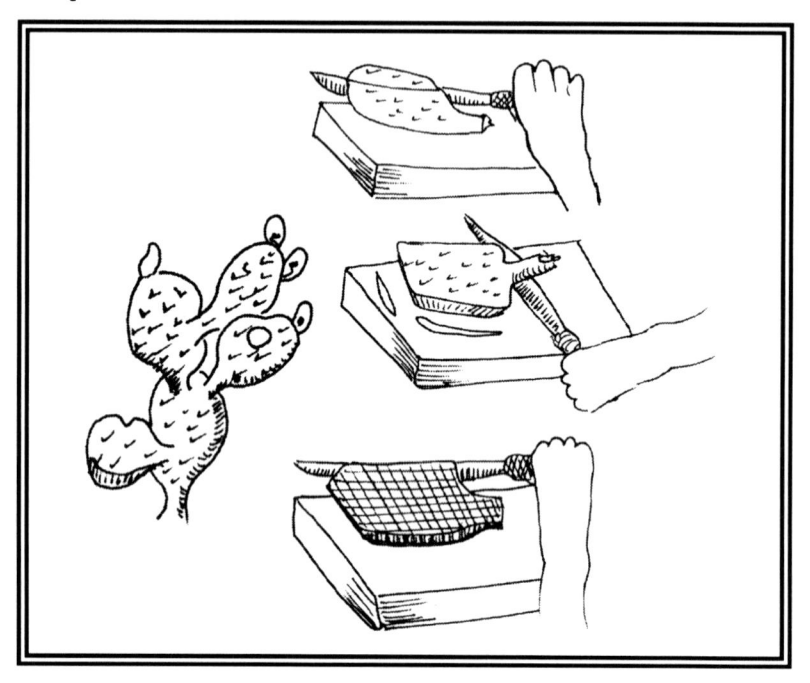

Preparing Prickly Pear.

Linda says- My personal attachment to this plant came in a way different from the usual. I led a wild food field trip to the desert in Black Canyon City. The temperature was only 90 degrees, but the sun was vicious to my fair, sensitive skin. I knew my sun block was not working so well. I scraped the needles off some prickly pear and cut the pad in half, applied the cactus meat to my sunburned face. The prickly pear and I became good friends from then on!

Purslane

Portulaca species
Purslane Family, *Portulacaceae*

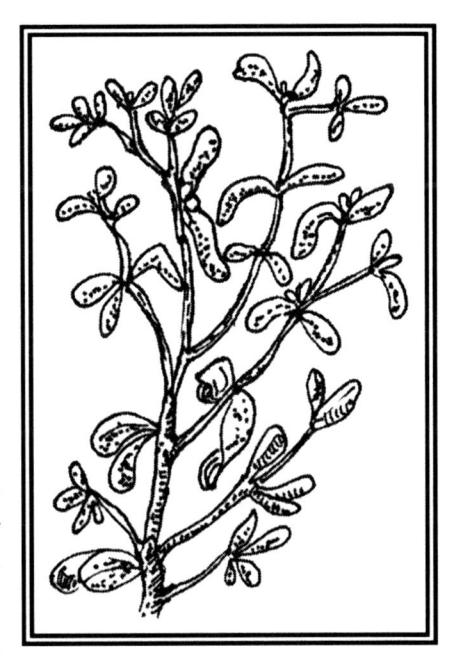

Other Names: Low pigweed, pusley.
History: Native to Europe, widely naturalized in United States.
Habitat: Fertile soil, topsoil.
Characteristics: Annual herb. Prostrate peddle-shaped succulent leaves on branching fleshy vines. Forms mats in masses of growth. Five to 7 petaled tiny yellow or red flowers are prolific. Multitudes of tiny black seeds produced on each plant.
Primary Uses: Culinary. All parts edible, raw or cooked. Purslane also pickled.
Nutritional Value: Shoots contain a balance of minerals for only 16 calories per half a cup.
Medical Value: A succulent that quenches thirst as well as puts a quantity of minerals in the body of an active person. Contains a high mineral content.
Collection and Storage: Gather mats, snapping off branches, and leave root for rejuvenation. Gather succulent young branches for raw treats. Also grows well indoors.

Linda says- My first harvest of purslane yielded 34 jars of pickles. Since then I have picked and pickled virtually hundreds of jars of purslane.

Queen Anne's Lace

Daucus carota
Parsley Family,
      Umbelliferae

Other Names: Wild Carrot.
History: Native to Europe, especially common in England; widespread in United States. Used by pioneers and Native Americans.
Habitat: Fields, waste areas.
Characteristics: Biennial herb. Reaches height of 2 feet or more. Carrot-like odor in stem and leaves as well as seeds. Flower clusters flat-topped, umbrella-like, lacy and often have single purple flower in the center. Old flowers "bird cage up" and turn brown. Leaves are parsley-like, feathery and smell like a carrot. All parts of stems have tiny fuzz, essential for proper identification.
Primary Uses: Culinary. Stems may be cut into sections and used for flavoring in stews & soups. Buds may be sautéed in oil.
Nutritional Value: Beta-carotene is exceptionally high, as is niacin.
Medicinal Value: The brown dried seeds are an excellent salt substitute.
Collection and Storage: Leaves are gathered in early spring and young leaves all season. Gather roots as carrots in early spring or fall; may be dug under leaves or hay even after winter snow. Gather flowers in summer. Collect seeds in fall.
Caution: Seeds of the wild carrot have a high content of vitamin A and shouldn't be eaten in excess.
Caution: Easily confused with Poison hemlock, resulting in severe burns of the fingers and hands as well as almost certain death if ingested. Check for the fuzz on the stem, as well as sandpaper-like feeling on dried stems from old dried fuzz. Use the foraging rule and crush, roll, and smell first; do not ingest. Determine the carrot smell and then look for the hair or fuzz.

Comments: In the fall, the flowers of wild carrot curl up; I call this "bird caging." At this time, I gather the seeds and a few seeds go a long way! I get a winter's supply of "salt" from a few flowers.

Queen Anne's Lace Chips

I will always remember my switch from potato chips to Queen Anne's lace flowers. In the wilderness, potato chips are a rare commodity, but Queen Anne's lace flowers crisp to a "chip" of carroty crispy flavor. I dip the flowers in hot oil and then put the brown crispy delights on hot rocks to cool. Todd loved to fly by and grab a few crispy flowers and pop them in his mouth using the stems as a handle.

Raspberry

Species
Rose Family, *Rosaceae*

Types: Cloudberry, Thimbleberry, wild red.
History: Some varieties native to North America, others naturalized from Europe, Asia. Used as a fruit in all countries where raspberries are found, especially Europe.
Habitat: Roadsides, rocky fields, thickets.
Characteristics: Hardy shrub, biennial canes from perennial roots. Varies in size to average of 5 feet high. Saw-toothed leaves in groups of 3. Round, erect stem with short thorns. Stems have a white powder on them. Flower white; fruit pebbly and red or purple when ripe, pulls away from receptacle.
Primary Uses: Culinary, medicinal, cosmetics. Early leaves used fresh or dried for tea. Buds and flowers are edible. Frozen in ice for drinks. Fruit eaten raw, cooked for jellies, jams, syrups, fresh for juices, wines. Stem peeled and chewed, tea made from young twigs.

Nutritional Value: High in vitamin C.
Medicinal Value: Tea from leaves good for control of diarrhea. Relaxes muscles of the uterus, beneficial for menstrual cramps and strengthening the reproductive system.
Cosmetic Value: Stimulating astringent in herbal hair rinses and bath mixtures.
Collection and Storage: For quick, efficient berry harvesting, hang a container around your waist and pick using both hands. When drying berries, spread a fine cloth over trays of berries to keep flies and birds away. Harvest young leaves.
Caution: Reddish-tinged mature leaves may be toxic. Use only young green leaves.

Linda says- The Adirondacks are inundated with black bears during berry season. One time I came upon a huge mama bear lying on her back under a thicket of raspberry. She was stupefied from gorging on the fermented, overripe berries. As she rolled over, she hiccuped and burped, then waddled off. See page vii.

Rose

Rosa rugosa, R. Caroline
Rose Family, *Rosaceae*

Other Names: Garden rose, wild rose.
History: R. Caroline native to North America; R. rugosa native to Asia but naturalized in United States. The symbol of love and beauty in ancient mythology. In Middle Ages, rose was symbol of secrecy (Subrosa). Also a heraldic symbol (War of Roses). Native Americans used all parts of plant. Europe and Asia (Sturtevant).
Habitat: Meadows, fields, woods, and coastal beaches.
Characteristics: Perennial or hardy shrub. Usually 5-petal flower,

yellow, pink, red or purple. Thorny stems, particularly central stem. Fruit is called <u>rose hip</u> and has prominent calyx lobes.

<u>Primary Uses</u>: Culinary, medicinal, cosmetic. Petals eaten raw or cooked, in salads, as a garnish. Petals are used for teas for rose ice cubes or popsicles. Rosehips remain at end of stem after the petals fall off. This bulbous end is edible raw or cooked, sliced in salads. Leaves used raw in salads; dried for leaf tea. Stems peeled and brewed as tea; roots peeled and chewed; brewed as tea. Hips used in jam, jellies, tea, candy, or confections. Brewed as tea; dry and reuse, or store in glass jars. Also used in incense & lamp oil.

Rose Hip.

<u>Nutritional Value</u>: Rose hips are high in vitamin C.

<u>Medicinal Value</u>: Known to help with colds and coughs. All parts, including roots, pounded and powdered as snuff to clear sinuses. In the past, the bark was mashed and applied to boils, wounds, sores. Today, rose oil is used to soothe headaches and as a healing ointment.

<u>Cosmetic Value</u>: Soaps & perfumes.

<u>Collection and Storage</u>: Gather all parts. Early rose shoots are easily clipped with scissors. Save all parts. Leaves may be clipped at any time and wear gloves and use a canvas bag or paper bag, one for leaves, one for roses and hips. Full roses snip 1 inch below hip. Pull petals off and use hips later.

Sauté rose petals in light oil for delicious treat. Roll in confectioner's sugar and refrigerate for candy.

Make rose oil by adding 1 ounce of rose powder (ground petals) to 4 tablespoons olive oil and store for a few weeks.

Linda says- Imagine what it's like to find yourself in a field of wild roses. Wandering among the bushes, I felt I was in the Garden of Eden. I wore gloves, dungarees and a long sleeve shirt for this experience. Nearby the Hudson River flowed past. Dragonflies and birds flew around me. When I tired of stuffing bags with petals for drying, I took a dip in the clean upper Hudson.

Saguaro

Carnegiea gigantea
Cactus Family, *Cactaceae*

<u>Other Names</u>: Monument of the desert.
<u>History</u>: Native to North American desert. Used by desert dwellers, Native Americans. Papago Indians make heavy syrup and intoxicating wines.
<u>Habitat</u>: Rocky, gravel soil; hills, canyons, and dry washes.
<u>Characteristics</u>: Treelike cactus. Has one or more founded arms extending branch-like from a single, thick trunk. Shallow root system with small roots radiating out the height of the cactus and then some; no taproot. Long thick trunk can be 2 1/2 feet wide and 50 feet high. Large, thick grooves running laterally on all parts; glochids and thick fishhook barbs on all parts of the cactus. Circle of white flowers on top branch. Fruit is egg shaped, 2 to 3 inches long.
<u>Primary Uses</u>: Culinary. Flower buds and seeds are eaten. Flowers are seen first, then buds with seeds inside. Fruit is eaten raw; husks and seeds are boiled, baked; fruits, husks and seeds used for syrup or jam. Dehydrated pulp used for flour, also oil, soft drinks, wine, and vinegar.
<u>Nutritional Value</u>: Buds, or fruits are high in niacin, fiber, carbohydrates as well as ash.
<u>Collection and Storage</u>: Harvest fruit in July and monthly thereafter by using a long stick to knock off bulbs. The skin splits and bark curls back when fruit is mature. May be dried for storage or be cut in half and soaked in water for about 1 hour after removing seeds.
<u>Caution</u>: Saguaro is a protected cactus tree. The Department of Agriculture must give permission to remove cactus from the desert. Fruits, buds may be used for food in a survival way, but permission must be noted in larger quantities. No permission will be granted to remove a rare crested specie of saguaro.

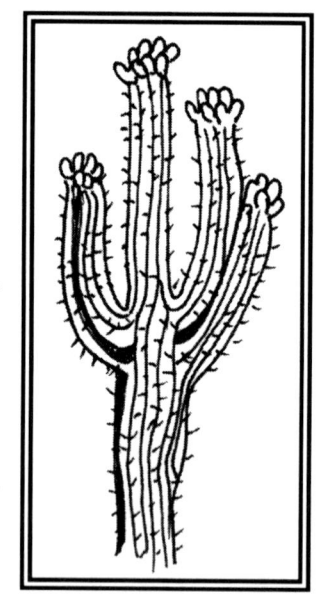

Linda says- The saguaro is a monument to desert life, protecting myriads of insects, mice, fruit-eating bats, owls, and other creatures of the desert. During a field walk with Willie Whitefeather, my Cherokee friend, I saw the buds of the giant saguaro harvested correctly. Willie fashioned the dried ribs of an old saguaro hulk together, lashed a hook-like end on the tip. The tool was 16 feet long and bent like a giant fishing rod in the desert wind. Deftly, he pulled the saguaro buds to the waiting crowd below.

"Desert Saguaros" by Paul Runyon, Linda's brother.

Sheep Sorrel

Rumex acetosella
Buckwheat Family *Polygonaceae*

Other Names: Garden sorrel, field sorrel, R. acetoea.
History: Native to Europe and Asia, naturalized throughout North America. Garden sorrel is cultivated for use as cooked green.
Habitat: Waste areas, roadsides.
Characteristics: Perennial herb. Reddish-tinged grassy tops. Mature seeds grow in spreading areas, standing out easily. The arrow shaped leaves are sour in taste.

Flowers are tiny and give way to seed quickly.
Primary Uses: Culinary. Leaves or stalks, eaten raw in salads or cooked as vegetable. Can be used in cold drinks.
Nutritional Value: High in vitamin C.
Medicinal Value: Juice has mild antiseptic effect (crush and place on wound as a poultice). Juice acts as laxative.
Collection and Storage: Use scissors and shear clumps of seeds easily, collecting in a bag or basket. Strip leaves off separately for soups, drinks. Seeds dried and stored in glass jar for use as spicy food additive.
Caution: All parts contain oxalic acid crystals. These may inhibit the absorption of calcium in the body. Place in boiling water for a minute or so to destroy the crystals.

> **Linda says-** I was amazed to taste the arrow head shaped leaves and find they were the same bitey, sour lemony taste as wood sorrel. (See page 137.)

Shepherd's Purse

Capsella bursa-pastoris
Mustard Family, *Brassicacae*

Other Names: Heart seed, field cabbage, purse seed, pick purse.
History: Native to Europe, widely naturalized in North America. Used by Native Americans.
Habitat: Waste areas, roadsides.
Characteristics: Annual herb. Erect plant from 3 inches to 1 1/2 feet high. Slender stem comes up from a spreading rosette of leaves formed flat on the ground. Deeply lobed leaves resemble wild lettuce or dandelion, but are more symmetrical. Has small white flower

with petals only 1/12 to 1/8 inch long, jutting out from the branches on a single stalk. Flowers give rise to seeds that resemble upside-down hearts.

Primary Uses: Culinary, medicinal. Basal leaves eaten in sandwiches, salads, soups, vegetable. Stalk is stir-fried, eaten raw. Buds and flowers eaten raw. Seeds are stir-fried and used as pepper, spice, seasoning. Roots are used fresh or dried as substitute for ginger or candied in syrup.

Nutritional Value: High in vitamin K, vegetable protein, potassium, calcium as well as beta-carotene and minerals.

Medicinal Value: Dried to use as blood-clotter and valuable taken in quantity for internal bleeding.[1]

Collection and Storage: Look for upside-down heart shaped seeds to distinguish plant from other wild mustards and pepper grasses. Gather leaves at any stage. Gather seeds when ripe. Entire plant may be gathered and chopped for vegetable, as well as dried for additive. Add seed pods to soups and stews for a pepper-like flavor.

---

[1] (Harrington 1967)

> **Linda says-** In the East I ate the little heart-shaped seedpods with gusto, but in the West I seek them with a vengeance. Shepherd's purse is known to help relieve strong menses and internal bleeding. According to Peter Bigfoot in his book *Arizona Wild Herbs,* pick the mature, flowering plants in the Arizona Desert foothills in springtime. Dosage: 1 teaspoon dried herb per cup of hot water. Steep 15 minutes and drink the tea cool. Take 2 or more cups full per day as needed. Drink it quickly, do not sip for best effect with internal bleeding.

Sow Thistle

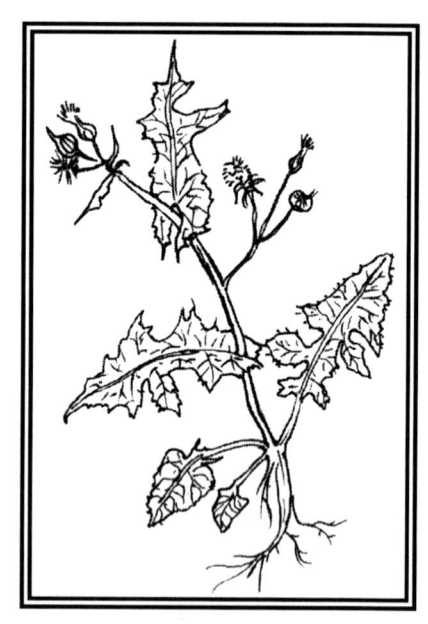

Sonchus oleraceus
Composite Family, *Compositae*

Other Names: Hoitgamivakhi (Pima Indian).
History: Native to North Africa, Europe, and western Asia; introduced to the United States and now found throughout most of Arizona. Used by Native Americans as greens.
Habitat: Roadsides, damp fields, waste areas.
Characteristics: Annual herb. Dandelion-type deeply lobed leaves begin from a central stem on the ground, clasping around stalk to a sharp point, like a sow's ear. Top of stalk bears branches, at the end of each grows a yellow flower that looks somewhat like a dandelion flower. After seeding, a white fuzz appears with single seeds on the end of the fluff.
Primary Uses: Culinary. Buds, flowers, and seeds edible raw or cooked. Stems may be cut into inch long pieces with scissors and stir fried or sautéed in oil as beans. Roots may be scrubbed and stewed in crock pot. Young leaves gathered and eaten as spring green, also eaten raw. Flowers stir-fried as vegetable or dried as well as frozen.
Nutritional Value: Very high in minerals.
Medical Uses: Milk reported to be somewhat diuretic.
Collection and Storage: One may trim leaves off long stems easily by leaving plant in ground and trimming off with a long knife. Use long handled knife or scissors, and wear gloves. Cut leaves over a basket or bag. To harvest large amounts, cut 2 long goldenrod or wild lettuce stalks and lay parallel to each other. Using a gloved hand, place the thistle plants across the parallel stalks. When a bundle is accumulated, pull golden rod stalks up and wrap around. Carry by holding both ends of the goldenrod together.

Linda says- My introduction to sow thistle was an experience. Early in the morning, when the dew was still on the grass, I wandered through a field of barley grass. The sow thistle was a succulent emerald green amid the grasses. Since I carry water with me, I rinsed off the leaves and munched on the sweetest breakfast I ever had in the West.

Strawberry

Fragaria virginiana
Rose Family, *Rosaceae*

Other Names:
Heartberry, wood strawberry.

History: Native to North America. Used in early European cultures, especially popular in France. Grows throughout Northern Hemisphere, excluding tropics. Native Americans steeped wild strawberries in warm water, then strained and cooled the liquid. The resulting lotion was a wash for irritations and skin ulcers.

Habitat: Fields, roadsides, meadows.

Characteristics: Low perennial, reaching height of 2 inches to 8 inches. Lobed saw-toothed basal leaves in groups of 3. Tiny, heart-shaped strawberry fruits. Small white flowers.

Primary Uses: Culinary, medicinal, cosmetic. Flowers are edible raw. Leaves eaten raw, dried for tea. Fruit eaten raw; cooked for jellies, jams, syrups; raw for juices, wines. Entire plant is steamed, boiled, used in soups and stews; cooking liquid is drunk.

Nutritional Value: Leaves and fruit high in vitamin C.

Medicinal Value: Leaves infused and used for sore throats. Juice is mixed with water for eye wash. Historically, roots are infused for gonorrhea remedy. Fruits are dried for stomach problems, diarrhea, dysentery, liver pain, jaundice. To juice, put a nylon stocking over a bowl. Place washed berries in stocking, then squeeze the juice into the bowl.

Cosmetic Value: Astringent and facial cleanser, useful for oily skin. Fruit is tartar remover for teeth; crush fruit, rub on, and rinse.

Collection and Storage: Pick fruits as you would cultivated strawberries. With container around your waist, pick fruit and leaves easily using both hands.
Caution: Dry leaves thoroughly for tea. Partially dried leaves can cause a toxic reaction.

> **Linda says-** Picking wild strawberries in the Adirondacks is difficult because the berries are very small and they ripen in blackfly season. The hardiest folk pick them in the wee morning hours or at night or in a rain. Snapping turtles eat the berries, as do other critters. I enjoy finding berries with chunks missing, and speculate what had breakfasted before me—insect, turtle, or mouse.

Sumac

Rhus species
Cashew Family, *Anacardiaceae*

Other Names: Staghorn, smooth, scarlet, dwarf, shining, mountain, hairy, velvet, fragrant, virginian, and winged sumacs; lemonade tree.
History: Mostly native to North America. Used by Native Americans and colonists.
Habitat: Roadsides, field edges.
Characteristics: Deciduous tree or shrub. Bushes or shrubs with some varieties to height of 20 feet. Fern-like leaves 14 to 24 inches long. Flaming red fuzzy berry clusters, hairy and lemonade tasting. Stems and branches exude white milk when broken. Plants need heavy sunshine to produce berry clusters.

Primary Uses: Culinary. Flowers and seeds are the fruits. Fruits eaten raw or crushed in water as lemonade. Frozen in ice cube trays for popsicles.
Nutritional Value: High in vitamin C.
Collection and Storage: Collect fruit before completely mature, when spires turn dark red. Collect spires by breaking off stems. For tea, strain out hairs by overlapping balsam branches as a sieve or use cheesecloth. Look carefully at the whole tree to avoid accidental contact. I once pulled down a poison sumac branch growing in the midst of an edible sumac tree.

Caution: Poison sumac has white or greenish-brown clusters instead of red spires. When dried, a tan spire is evident. See Part V.

> ***Linda says-*** There is nothing more beautiful than a glass decanter filled with sumac berries. Scarlet tea! Sumac lemonade is a year-round treat. Collect and rinse the heads early in the season. Do not wait until fall, when insects have made their nests in them! Summer seed heads may be dried and stored in glass, or bagged and frozen.

Sunflower

Helianthus species
Composite Family, *Compositae*

Types: Annual, wild, Western, common, tall, giant, weak, saw-toothed, swamp, little, aspen, showy, and woodland sunflower; Marigold of Peru.
History: Native to the Americas, probably originated in Peru. Used by Native Americans.
Habitat: Widespread but more common in West. Gardens, fields, sunny open areas.

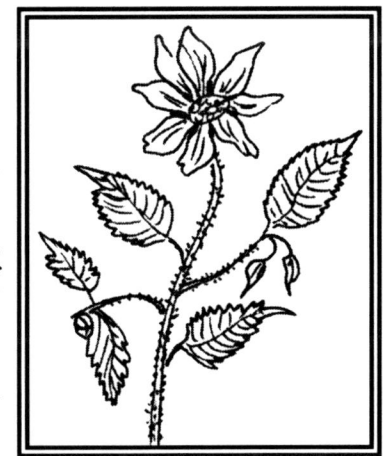

Characteristic: Annual herb. Composite flowers on plants from 3 to 14 feet high. Yellow ray flowers with purplish red to brown disk flowers in center. Stems may be smooth or hairy. Leaves opposite or alternate, characteristically narrow, long, rough, and saw-toothed.
Primary Uses: Culinary, cosmetic. Seeds crushed and boiled; oil skimmed off top and eaten. Pith edible raw and cooked in soups. Used for paper making, organic fertilizer, oil, lubricants, candles. Petals used for yellow dye.
Nutritional Value: Seeds high in protein.
Cosmetic Value: Oil, soaps.
Collection and Storage: After flower matures and petals fall off, gather seeds. Snap off dried center and dry seeds on screen, then bottle. Large heads dried in a closed paper bag will mature eventually. Scrape the heads across a sieve with holes sized such that the seeds can drop through. Place seeds on cookie sheet and roast in dry, low oven. Bottle in glass for long-term storage. Pith is collected by peeling outer fresh stem when plant is dry and pulling out white pith.

Linda says- Snapping off the tiny dried heads of desert sunflowers is fun and very profitable for my table. A few crushed heads and a bit of water yielded a nutritious gruel with an obvious energy boost. Storage is simple: I put the dried sunflower heads in a glass jar.

Thistle

Circium species
Composite Family, *Compositae*

Other Names: Bull thistle, yellow thistle, plumed thistle, swamp thistle, field thistle.
History: Native to North America. Used since colonial times and by Native Americans.

Habitat: Roadside, fields, waste areas.
Characteristics: Biennial plant with dozens of varieties. Reaches height of 2 to 6 feet. Leaves and stems have sharp spines (bulbs) on all parts. One to 3 large, purplish flower heads with spines around base produced second year.
Primary Uses: Culinary. Flowers steamed and eaten raw much like an artichoke. Leaves dried and ground to a fine, green flour. Dried leaves also used for tea. Steamed leaves eaten as a vegetable. Stems peeled, sliced, and cooked as a vegetable. Roots boiled for tea.
Nutritional Value: High in potassium, phosphorus, and vegetable protein.
Medicinal Value: Roots are used for restorative tonic. Roots can be harvested year round, but early summer is best.
Collection and Storage: Early in the spring, use a penknife to cut beneath the basal leaves as you would a head of lettuce, being careful not to disturb the root. Cut leaves off central stem, holding a bag underneath. Use tongs and strainer to wash. Dries and freezes well.

> **Linda says-** The beginnings of a thistle plant are succulent. Its thorny whorls are clear blue-green. The trick is to get as much of the core as possible, leaving the basal root. The thistle will grow again if you carefully take the upper part of the root along with the leaves, leaving the deeper root alone. The first time I did this I wondered if it was worth the effort. Now I use tongs and a strainer, and find the preparation time negligible while the taste is lasting!

Thyme

Thymus vulgaris
Mint Family, *Labiatae*

Other Names: Bee plant, wild thyme, mountain thyme, creeping thyme.
History: Native to Europe; in North America gardens vari-

ety has expanded to woods and fields. Adapted and used by Native Americans.

Habitat: Fields, roadsides, roadbanks, lawns.

Characteristics: Woody perennial. Prostrate herb growing to height of about 4 inches. Several varieties, all with pepper-like smell. Many spikes of leaves and lavender spike flowers. Grows in thick patches.

Primary Uses: Culinary, medicinal, cosmetic. All parts edible. Ground to powder; twigs, leaves, and flowers simmered for stimulant tea. Added as seasoning to soups, stews, fish, meat dishes.

Nutritional Value: Has a high count of vitamin C, iron and niacin.

Medicinal Value: For headaches, stimulant, antiseptic, aromatic, carminative, diuretic, diaphoretic, emenogogic, and antispasmodic. Treatment for sore throats.

Cosmetic Value: Oil is aromatic, antiseptic, diaphoretic, stimulant, disinfectant, and used in cosmetic lotions. An effective deodorant, mouthwash, aftershave lotion, and wash.

Collection and Storage: Easily sheared with scissors. Mow thyme patches regularly for a winter supply. For best results, rub dried thyme through a screen to remove stems and twigs.

It's Thyme Time

Lying next to a patch of lavender thyme flowers and nibbling the pungent, peppery twigs and flowers soon became a favorite activity after a busy day. My favorite patch was near Cedar River Road in the Adirondacks. Those days were spent building structures, gathering large amounts of food, experimenting with storage and cooking methods, and just plain surviving. I would lay full length next to the bees working the thyme. The bees never minded, lifting off and moving as I raised my hand to pluck a stem to munch on. The thyme was a perfect stimulant. I returned to finish the evening renewed.

Tumbleweed

Salsola kali
Goosefoot Family,
    Chenopodiaceae

Other Names: Russian thistle.
History: Native to Russia, brought to United States in flaxseed 100 years ago.
Habitat: Open desert, fields, roadsides, hot dry country.
Characteristics: Annual herb. Reaches height of 2 feet in bushy and roundish intricately branched plant. Often reddish with ridged stems and many tiny ends. Late in season, plant breaks off at base and rolls around freely, dropping its many seeds.
Primary Uses: Culinary. Tips eaten cooked as vegetable or in soups and stir-fries.
Nutritional Value: High in calcium, vegetable protein, carbohydrates, fiber, potassium and *extremely* high in niacin.
Collection and Storage: Choose tumbleweed that does not have many buds and flowers, and harvest young seedlings whenever possible. If you don't have time to stand and clip the bush, remove a few branches and let wilt naturally. Clip ends that wilt the most; this ensures getting the most succulent ends. Drop succulent ends in a bowl, wash well, or soak overnight in a large bowl of water and a teaspoon of vinegar. The more you clip the more the plant grows to maturation, producing many more branches.
Caution: Do not ingest raw except in extreme emergency, chewing thoroughly.

Linda says- Making tumbleweed soup is simple. I find a tumbleweed without prickly seeds and flowers, and shear the succulent ends into a basket. Wash the succulent tips well, quickly cook them in boiling water, then pack half-cup quantities into soup bags and freeze. You'll have a month's worth of delicious soup.

Violets

Viola species
Violet Family, *Violaceae*

Other Names: Heart leaf, baby face, wild violet.
History: World wide distribution, including North America. Used by Native Americans.
Habitat: Rich, moist woods, swamp sides, stream banks, damp lawns.
Characteristics: Annual or perennial herb. Smooth, heart-shaped leaves with little scalloping on edges. Single stem. Flowers on single stem from leaf base; five-petaled, arising from a center spur, usually yellow. Lower petal has heavy veins and there is a bearded design. Colors vary from white or yellow to purple or blue.
Primary Uses: Culinary, medicinal, cosmetic. Flowers used as little "fruits", eaten raw and candied; also used in jam, jelly, wine. Buds, flowers, and leaves dried for tea. Leaves eaten raw in sandwiches or salads; cooked, as thickening agent. Roots of violets may be eaten raw or cooked in a crockpot. Stems may be snipped into 1 inch sections and added as "beans" to soups.
Nutritional Value: High in vitamins A and C.
Medicinal Value: Good for colds, flu. Eaten for dry, scratchy throat. Raw flowers eaten in quantity for varicose veins.
Cosmetic Value: An astringent; used in cold creams, oil lotions (infusion of flowers).
Collection and Storage: Clip violets easily with scissors, but use fingers when gathering leaves. Choose large leaves to wrap up tofu, rice, potatoes, beans.
Caution: Yellow wood violets may be cathartic; small, common lawn violets are not.

Linda says- Tiny faces peeping out of the dew-covered grass - violets are a favorite of mine, whether next to a stream in early spring or on the back lawn. The violet's face is a spring flag, heralding renewed life.
    Todd loved the little faces in his ice cream. I would lightly stir a handful of violets into his vanilla ice cream.

Wild Lettuce

Lactuca canadensis
Composite Family, *Compositae*

Other Names: Milk lettuce, chicory lettuce, horseweed.
History: Native to North America. Used by Native Americans; The Menominees use wild lettuce as a sedative.[2]
Habitat: Fields, waste ground, roadsides.
Characteristics: Biennial herb. Deeply serrated leaves, coming off a central stalk. Tiny buds and extensions off of top stalk. Dozens of yellow dandelion-like ray flowers. White milk in stem. Stem veins are characteristic.
Primary Uses: Leaves eaten in salads. Flowers and buds are stir-fried. Flowers and seeds are eaten raw or cooked, sautéed in oil and garlic. Stems (young) cut to bean size and cooked. Leaves dried and ground for seasoning. Roots have been used as a gum.
Nutritional Value: High in beta-carotene, ascorbic acid, (vitamin C)
Medicinal Value: Tea from milky leaves is sedative. Sap induces sweating and is used as a fever reducer. Decoctions of sap are known to ease irritability and insomnia.
Collection and Storage: Collect leaves, stems, flowers as you would lettuce, use raw.

Linda says- The tops or buds of this great plant are delicious. Stir-fried with a little olive oil and garlic salt, wild lettuce makes a wild meal!

---

[2]Densmore, Francis. *How the Indians Use Wild Plants for Food.* New York: Dover, 1974.

Willow

Salix species
Willow Family, *Salicaceae*

Types: Beaked, black, blue, Canada, coastal plain, crack, Drummond's, ward, weeping willow.

History: Grows throughout the world but abundant in cooler parts of Northern Hemisphere. In North America, used by Eskimos, Native Americans. Eskimos dry inner bark, strip, and cook like spaghetti. Inner bark layer called keeleeyuk in Eskimo, meaning "the scrape."

Habitat: Damp areas, stream sides, swamps.

Characteristics: Deciduous shrubby tree. Over 100 species, many with drooping branches alternate 2 inch leaves, sectioned twig tendril. Flowers (catkins) are furry, present before leaves. Roots seek water; have thousands of white hair-like extensions.

Primary Uses: Culinary, medicinal, cosmetic. Inner bark eaten raw; may be dried and pounded to flour. Buds, flowers or catkins, seeds are brewed as tea. Leaves are emergency food, tea, but use sparingly as they contain "aspirin". Willow sprouts provide early spring food.

Nutritional Value: Very high in beta-carotene.

Medicinal Value: All parts have acetylsalicylic acid, or salix, an aspirin source. Ten to 15 catkins or one 6 to 8 inch tendril, equal approximately 2 aspirins.

Cosmetic Value: Astringent; used in lotions, creams, facial washes.

Collection and Storage: Break tendrils into 6-inch pieces and dry. Store in glass container. Collect catkins or leaves by running hand down tendril. Dry and store. A supply of aspirin was easily kept through the year by stripping catkins in the spring, drying, and placing in glass.

> **Beneath the Weeping Willow**
> I sit under my favorite weeping willow tree. My fingers fly as I make a supple willow basket. At the same time, the tiny catkins hanging from the twigs are irresistible. I first munch only four or five, remembering that these catkins are like aspirin. But the catkins are so delicious. I quickly down about 1/2 cup of them while I make the basket. When I stand up, my ears ring and I wobble, almost faint. I sit down immediately and put my head between my knees to stop the dizziness. I curl up and go to sleep. Much later, I awake with a gigantic headache that lasts almost two days. A hard-learned lesson about the potency of willow catkins!

Wintergreen

Gaultheria procumbens
Heath Family, *Ericaceae*

Other Names: Checkerberry.
History: Native to North America, especially East.
Habitat: Wooded areas, acid soils, under softwood trees.
Characteristics: Perennial herb. Drooping waxy white flowers hidden by flat shiny evergreen leaves. Has bright red berries.
Primary Uses: Culinary, medicinal, cosmetic. Leaves used fresh or dried, crush or steep for tea. Oils are distilled for flavorings. Buds, flowers, and seeds (berries) edible.
Nutritional Value: Fruit or berry, high in acetylsalicylic acid, or aspirin. Contains niacin.
Medicinal Value: Leaves contain acetylsalicylic acid (aspirin). Steeped for astringent. Boiled and applied to wounds or used as eyewash. Tea breaks a fever. Used as an anti-rheumatism gargle, to treat wounds or hemorrhages, as a poultice for insect bites and

bruises, or as vaginal douche (see page 138). Oils used for rheumatism or arthritis, sciatica.

Cosmetic Value: Aromatic oil used as flavoring for candies, toothpaste.

Collection and Storage: Snip leaves; dig roots, and collect berries. Wintergreen is known to grow under the snow; dig and eat during winter. Dried, powdered wintergreen leaves have more flavor than fresh leaves. Plant transplants easily, loves acid soil.

Caution: Contains acetylsalicylic acid (aspirin). Overdose of oil may produce drowsiness, congestion, and delirium.

> **Linda Says-** Driving from Plainfield, New Jersey, to Indian Lake in upstate New York was a 17-hour drive in 1940. My frustration as a child was greatly relieved by chewing a wintergreen leaf I picked in the woods. I soon learned to recognize the flat, shiny, three-leafed plant growing close to the ground. If I was lucky enough to find a red wintergreen berry, I savored that berry for an hour of the drive!

Winter Harvest

It was the first time I trooped through snowdrifts looking for food. I thought it would be hard work, but balsam and pine trees were always within reach of the trail. Bushes of meadowsweet poked above the snow. I would never starve if I ate twigs and bark. But I did not know then that the Iroquois Abanake (bark eaters) did just that for centuries. Following the deer trail, I located the wintergreen. Reaching deep into the snow opening, I pulled out a few plants. The shiny green aromatic leaf never tasted so good!

Wood Sorrel

Oxalis species
Wood-Sorrel Family,
Oxalidaceae

Other Names: Clover sorrel, shamrock sorrel, lemon sorrel.
History: Eaten as spicy salad for centuries. Introduced to North America and used by Europeans and all Native American cultures except Crow and Menomini. George Washington Carver, of peanut fame, made pies and confections from wood sorrel; he even made a paint from sorrel.
Habitat: Fields, woods, moist places, under trees, or shrubs.
Characteristics: Low-growing perennial herb. Leaves alternate or basal, in 3 heart-shaped leaflets (a shamrock), sour lemon tasting. Grows from 2 inches to 8 inches in bunches or carpets. Flowers long, with 5 yellow, rarely green petals, sometimes red at the base. Leaves fold up sometimes when extremely warm or in sunshine.
Primary Uses: Culinary, cosmetic. Leaves cooked in soups, stews. Seeds cooked in soups, stews. Buds and flowers cooked in soups, stews. Many commercial uses. Stems are edible. Cut into inch pieces and add to soups, stews.
Nutritional Value: High in vitamin C. Because wood sorrel is high in oxalic acid, it hastens the elimination of calcium from the body and inhibits absorption of calcium. Therefore, limit intake of raw leaves to no more than 15 at most. Cook for safety. Drop entire plant in hot water. When color changes from bright emerald green to olive green, oxalic acid is neutralized.
Cosmetic Value: Antiseptic wash for skin and skin eruptions.
Collection and Storage: Gather seeds when they are red. Gather leaves year-round. Collect flowers any time of year. Cut stems, leaves, and seeds to dry for soup or use as a salad ingredient.

Yarrow

Achillea millefolium
Composite Family, *Compositae*

Other Names: Milfoil, field hop, old man's pepper.
History: Native to Europe. Naturalized in North America.
Habitat: Old fields, roadsides.
Characteristics: Perennial herb. Grown to height of 2 1/2 feet. Creeping rootstock; fern-like, lacy leaves, slightly hairy. Flowers typically white in wild, but occur in red, rosy, yellow, and magenta in ornamental gardens. Flowers have 5 petal-like rays, dozens in flat top clusters. Each flower is on its own stem, originating from one stem.
Primary Uses: Culinary, medicinal, cosmetic.
Stimulant tea. Boil any part of plant in small amounts. Leaves used raw in salads, or cooked for green. Seeds and flowers crumpled for seasoning. First-year roots peeled and cooked as vegetable.
Nutritional Value: High in calcium, and potassium.
Medicinal Value: Moves fluids in the body. Diaphoretic; cleanses the blood; tonic, stimulant, antiseptic; soothes burns. Strengthens immune system.
Cosmetic Value: Astringent; cleansing herb in facial mixtures, vaginal douches, shampoos; for dandruff and baldness treatments. Mouthwash for toothaches.
Collection and Storage: Clip flowers with scissors or break off with fingers. Pull leaves down stalk.

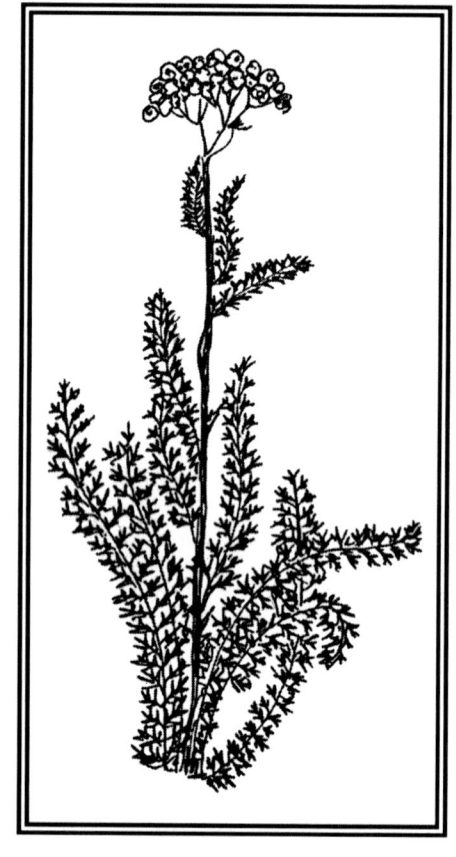

Yarrow for Teeth Cleansing

A dentist appointment loomed ahead. I hated to leave the forest and drive the old truck to Indian Lake, but having no phone, I could not cancel. I picked a yarrow stem and stripped off the leaves. I placed a leaf under my forefinger and scrubbed my teeth with the yarrow. Very astringent and satisfying. Using the stem, I picked here and there. Then I drove the pickup to town. In the dentist chair, the doctor exclaimed, "Linda, you've been in the yarrow again!" I did not realize that the yarrow had dyed the inside of my mouth green, especially my teeth—temporarily, of course.

The author begins her exploration of wild food.

Notes

Aloe Vera
Aloe perfoliata vera Cactus Family, Liliaceae

Amaranth
Amaranthus retroflexus Amaranth Family, Amaranthaceae

Arrowhead
Sagittaria* species, *S. chinensis Water Plantain Family, Alismataceae

Aster
Aster nemoralis Composite Family, Asteraceae

Balsam Fir
Abies balsamea *Pine Family, Pinaceae*

Birch
Betula species *Birch Family, Betulaceae*

Blackberry
Rubus villosus and other bramble berry species Rose Family, Rosaceae

Blueberry
Vaccinium myrtillus Heath Family, Erraceae

Bulrush
Scirpus validus Sedge Family, Juncaceae

Burdock
Arctium lappa Sunflower Family, Asteraceae

Cattail
Typha latifolia   Cattail Family, Typhacerae

Chamomile
Matricaria chamomilla   Composite Family, Asteraceae

Chickweed
Stellaria media (common chickweed), Pink Family, Caryphyllea

Chicory
Cichorium intybus Composite Family, Asteraceae

Cholla
Opuntia fulgida   *Cactus Family, Cactaceae*

Clover, red
Trifolium pratense   *Legume Family, Leguminosea*

Clover, white
Trefolium Repens *Legume Family, Leguminosae*

Crabgrass
Digitaria sanguinalis *Grass Family, Graminea*

Daisy
Chrysanthemum leucanthemum *Composite Family, Asteraceae*

Dandelion
Taraxacum officinale *Composite Family, Asteraceae*

Dock
Rumex crispus Buckwheat Family, Portulaceae

Evening Primrose
Oenothera biennis Evening Primrose Family, Onagraceae

Filarie
Erodium circutarium *Cranesbill Family, Geraniaceae*

Fireweed
Epilobium angustifolium *Evening Primrose Family, Onagraceae*

Goldenrod
Solidago odora Sunflower Family, Compositae

Grape
***Vitis arizonica* (Western) *Vitis labrusca* (Eastern) *Vitis rotundifolia* (South)**
Vitis Family, Vitaceae

Lamb's Quarters
Chenopodium album *Goosefoot Family, Chenopodiaceae*

Malva
Malva Neglecta *Mallow Family, Malvaceae*

Maple
Aceraceae Family

Meadowsweet
Filipendula ulmaria   Rose Family, Rosaceae

Milk Thistle
Silybum marianum   Composite Family, Compositae

Milkweed
Asclepias species   Milkweed Family, Aslepiadaceae

Mint
Mentha species *Labiatae Family*

Mullein
Verbascum thapsus *Snapdragon Family, Scrophulariaceae*

Mustard, yellow
Brassica species Mustard Family, Brassiacaceae

Mustard, black
Brassica species Mustard Family, Brassicerceae

Nettles
Urtica species *Urticaceae Family*

Phragmities
Phragmities communis *Grass Family, Graminae*

Pine
Pinus species *Pine Family, Pinaceae*

Plantain
Plantago major (common plantain) *Plantain Family, Plantaginaceae*

Prickly Pear
Opuntia megacantha *Cactus Family, Cactaceae*

Purslane
Portulaca species *Purslane Family, Portulacaceae*

Queen Anne's Lace
Daucus carota   Parsley Family, Umbelliferae

Raspberry
Species Rose Family, Rosaceae

Rose
Rosa rugosa, R. Caroline   Rose Family, Rosaceae

Saguaro
Carnegiea gigantea   Cactus Family, Cactaceae

Sheep Sorrel
Rumex acetosella *Buckwheat Family, Polygonaceae*

Shepherd's Purse
Capsella bursa-pastoris *Mustard Family, Brassicaceae*

Sow Thistle
Sonchus oleraceus *Composite Family, Compositae*

Strawberry
Fragaria virginiana *Rose Family, Rosaceae*

Sumac
Rhus species *Cashew Family, Anacardiaceae*

Sunflower
Helianthus species *Composite Family, Compositae*

Thistle
Circium species *Composite Family, Compositae*

Thyme
Thymus vulgaris *Mint Family, Labiatae*

Tumbleweed
Salsola kali   Goosefoot Family, Chenopodiaceae

Violets
Viola species   Violet Family, Violaceae

Wild Lettuce
Lactuca canadensis Composite Family, Compositae

Willow
Salix species Willow Family, Salicaceae

Wintergreen
Gaultheria procumbens Heath Family, Ericaceae

Wood Sorrel
Oxalis species Wood Sorrel Family, Oxalidaceae

Yarrow
Achillea millefolium *Composite Family, Compositae*

Linda building a walk.

Field in Pots

Home Health Aides Provided Lunch for Linda

Nutrient Value of Wild Foods

I do not advocate giving up one's daily diet, but rather supplementing it with wild foods. My homesteading days taught me the value of a balanced vegetarian diet. Note the following dated 1997 - Some variations have been noted to date.

Some USDA recommended daily dietary allowances for adults are:

- Vitamin A—1,000 micrograms (males) and 800 micrograms (females)
- Vitamin C—60 milligrams
- Vitamin D—5 micrograms
- Vitamin B1 (thiamin)—1.4 milligrams (males) or 1.0 milligrams (females)
- Vitamin B2 (riboflavin)—1.6 milligrams (males) or 1.2 milligrams (females)
- Vitamin B3—20 milligrams
- Vitamin B6—2 milligrams
- Vitamin B12—3 micrograms
- Calcium—1,200 milligrams
- Iron—18 milligrams
- Phosphorous—800 milligrams
- Potassium—1,875 milligrams
- Protein—56 grams (males) or 44 grams (females)

So, you can see from the above how well wild foods can boost the nutritional value of your meal.

The listings that follow show the nutritional values for 1/2 cup quantities of uncooked plant matter.

As much as possible, the data was derived from Duke and Atchley, 1986 *CRC Handbook of Proximate Analyses*, CRC Press, 1986. Duke's FNF database; or from Joseph Laferriere's Nutricomp database. When no other data were available we used Mark Pedersen's Nutritional Herbology (Pedersen Publishing, 1987).

When available data was not specific to the plant or plant part we used data from a related species, or genus, or similar species in a different family. When absolutely no pertinent data was uncovered—as, for example, for violet flowers—averages for flowers such as those published by Duke and Atchley, were used. Occasionally it was necessary to estimate the water content, back-calculating from dried to a fresh basis. Adjustments were then made for carbohydrate and/or fiber content, ensuring that the figures were within a normal range for the plant.

How to read the Nutritional Listings

All quantities given are amounts per 1/2 cup raw food item (≈100 grams).

The figure for Calories is number of Calories per 1/2 cup (100 g) raw food.

Column 1: Measurements are in grams (g) per 100 g (1/2 cup). (Because of the mathematical coincidence, this could also be expressed as a percentage of the whole.)
Abbreviations:
- Carbos. = Carbohydrates
- Tr. = trace amount.

Column 2: Measurements are in milligrams (mg) per 100 g (1/2 cup).
Abbreviations:
- Phos. = Phosphorus
- Ascorbic. = Ascorbic acid
- Tr. = trace amount.

Column 3: Measurements are in micrograms (µg) per 100 g (1/2 cup).
Abbreviations:
- β Carotene = Beta-carotene
- Tr. = trace amount.

No. 1 Aloe Juice

Amount of nutrient per 100g. (About ½ cup)				Calories: 1	
Water	99.5 g	Calcium	2 mg	β Carotene	Tr.
Protein	0.1 g	Phos.	2 mg	Thiamin	Tr.
Fat	Tr.	Iron	0.2 mg	Riboflavin	Tr.
Carbos.	0.4 g	Sodium	0.3 mg	Niacin	Tr.
Fiber	0.1 g	Potassium	0.4 mg		
Ash	Tr.	Ascorbic.	3.1 mg		

No. 2 Amaranth Leaves

Amount of nutrient per 100g. (About ½ cup)				Calories: 35	
Water	91.6 g	Calcium	448 mg	β Carotene	4300 μg
Protein	2.9 g	Phos.	85 mg	Thiamin	65 μg
Fat	0.4 g	Iron	13 mg	Riboflavin	300 μg
Carbos.	4.5 g	Sodium	20 mg	Niacin	1300 μg
Fiber	1.1 g	Potassium	617 mg		
Ash	1.5 g	Ascorbic.	53 mg		

No. 3 Amaranth Seed

Amount of nutrient per 100g. (About ½ cup)				Calories: 358	
Water	12.3 g	Calcium	247 mg	β Carotene	0.0 μg
Protein	12.9 g	Phos.	500 mg	Thiamin	140 μg
Fat	7.2 g	Iron	3.4 mg	Riboflavin	320 μg
Carbos.	65.1 g	Sodium	33.5 mg	Niacin	1000 μg
Fiber	6.7 g	Potassium	52.5 mg		
Ash	2.5 g	Ascorbic.	3.0 mg		

No. 4 Arrowhead Tubers

Amount of nutrient per 100g. (About ½ cup)				Calories: 99	
Water	72.5 g	Calcium	10 mg	β Carotene	0 μg
Protein	5.3 g	Phos.	174 mg	Thiamin	200 μg
Fat	0.3 g	Iron	2.6 mg	Riboflavin	100 μg
Carbos.	20.2 g	Sodium	225 mg	Niacin	1700 μg
Fiber	0.8 g	Potassium	922 mg		
Ash	1.7 g	Ascorbic.	1 mg		

No. 5 Arrowhead Shoots

Amount of nutrient per 100g. (About ½ cup)				Calories: 62	
Water	85 g	Calcium	82 mg	β Carotene	1700 μg
Protein	2.6 g	Phos.	45 mg	Thiamin	200 μg
Fat	1.0 g	Iron	7 mg	Riboflavin	300 μg
Carbos.	9.9 g	Sodium	21 mg	Niacin	1500 μg
Fiber	0.9 g	Potassium	606 mg		
Ash	1.5 g	Ascorbic.	60 mg		

No. 6 Aster Leaves

Amount of nutrient per 100g. (About ½ cup)				Calories: 39	
Water	87.2 g	Calcium	42 mg	β Carotene	3345 μg
Protein	4.2 g	Phos.	76 mg	Thiamin	180 μg
Fat	0.7 g	Iron	4 mg	Riboflavin	360 μg
Carbos.	6.4 g	Sodium	40 mg	Niacin	1100 μg
Fiber	1.1 g	Potassium	533 mg		
Ash	1.5 g	Ascorbic.	88 mg		

No. 7 Balsam Fir Needles

Amount of nutrient per 100g. (About ½ cup)				Calories: 35	
Water	88 g	Calcium	186 mg	β Carotene	5652 μg
Protein	2.8 g	Phos.	57 mg	Thiamin	110 μg
Fat	0.5 g	Iron	2.7 mg	Riboflavin	240 μg
Carbos.	6.5 g	Sodium	40 mg	Niacin	900 μg
Fiber	1.3 g	Potassium	382 mg		
Ash	2.1 g	Ascorbic.	64 mg		

No. 8 Birch Leaves

Amount of nutrient per 100g. (About ½ cup)				Calories: 1	
Water	88.0 g	Calcium	186 mg	β Carotene	5652 μg
Protein	3.3 g	Phos.	57 mg	Thiamin	110 μg
Fat	1.0 g	Iron	2.7 mg	Riboflavin	240 μg
Carbos.	6.7 g	Sodium	40 mg	Niacin	900 μg
Fiber	2.0 g	Potassium	382 mg		
Ash	0.9 g	Ascorbic.	68 mg		

No. 9 Blackberry Fruit

Amount of nutrient per 100g. (About ½ cup)				Calories: 57	
Water	84.4 g	Calcium	29 mg	β Carotene	1000 µg
Protein	1.2 g	Phos.	26 mg	Thiamin	200 µg
Fat	0.7 g	Iron	1.3 mg	Riboflavin	500 µg
Carbos.	13.3 g	Sodium	1 mg	Niacin	600 µg
Fiber	3.7 g	Potassium	168 mg		
Ash	0.5 g	Ascorbic.	22 mg		

No. 10 Blueberry Fruit

Amount of nutrient per 100g. (About ½ cup)				Calories: 62	
Water	83.2 g	Calcium	15 mg	β Carotene	60 µg
Protein	0.7 g	Phos.	13 mg	Thiamin	30 µg
Fat	0.5 g	Iron	1 mg	Riboflavin	60 µg
Carbos.	15.3 g	Sodium	1 mg	Niacin	500 µg
Fiber	1.5 g	Potassium	81 mg		
Ash	0.3 g	Ascorbic.	14 mg		

No. 11 Bulrush Flowers

Amount of nutrient per 100g. (About ½ cup)				Calories: 16	
Water	94.8 g	Calcium	47 mg	β Carotene	400 µg
Protein	0.4 g	Phos.	86 mg	Thiamin	20 µg
Fat	0.1 g	Iron	1.0 mg	Riboflavin	110 µg
Carbos.	4.4 g	Sodium	18 mg	Niacin	600 µg
Fiber	1.2 g	Potassium	166 mg		
Ash	0.3 g	Ascorbic.	18 mg		

No. 12 Bulrush Seeds

Amount of nutrient per 100g. (About ½ cup)				Calories: 342	
Water	11.1 g	Calcium	24 mg	β Carotene	0 µg
Protein	10.7 g	Phos.	267 mg	Thiamin	450 µg
Fat	8.9 g	Iron	3.1 mg	Riboflavin	200 µg
Carbos.	67.7 g	Sodium	6 mg	Niacin	3800 µg
Fiber	2.6 g	Potassium	316 mg		
Ash	1.6 g	Ascorbic.	0 mg		

No. 13 Bulrush Shoots

Amount of nutrient per 100g. (About ½ cup)				Calories: 31	
Water	89.9 g	Calcium	50 mg	β Carotene	1166 µg
Protein	0.8 g	Phos.	18 mg	Thiamin	150 µg
Fat	0.2 g	Iron	1.1 mg	Riboflavin	190 µg
Carbos.	7.5 g	Sodium	9 mg	Niacin	1000 µg
Fiber	2.8 g	Potassium	283 mg		
Ash	1.4 g	Ascorbic.	41 mg		

No. 14 Burdock Root

Amount of nutrient per 100g. (About ½ cup)				Calories: 89	
Water	76.5 g	Calcium	50 mg	β Carotene	0 µg
Protein	2.5 g	Phos.	58 mg	Thiamin	250 µg
Fat	0.1 g	Iron	1.2 mg	Riboflavin	80 µg
Carbos.	20.1 g	Sodium	30 mg	Niacin	300 µg
Fiber	1.7 g	Potassium	180 mg		
Ash	0.8 g	Ascorbic.	2 mg		

No. 15 Cattail Roots

Amount of nutrient per 100g. (About ½ cup)				Calories: 73	
Water	79.7 g	Calcium	32 mg	β Carotene	568 µg
Protein	1.6 g	Phos.	54 mg	Thiamin	90 µg
Fat	0.6 g	Iron	1 mg	Riboflavin	50 µg
Carbos.	17.5 g	Sodium	14 mg	Niacin	800 µg
Fiber	1.0 g	Potassium	399 mg		
Ash	1.0 g	Ascorbic.	18 mg		

No. 16 Cattail Shoots

Amount of nutrient per 100g. (About ½ cup)				Calories: 31	
Water	81.3 g	Calcium	58 mg	β Carotene	2200 µg
Protein	1.8 g	Phos.	109 mg	Thiamin	300 µg
Fat	0.3 g	Iron	2 mg	Riboflavin	400 µg
Carbos.	15.0 g	Sodium	8 mg	Niacin	800 µg
Fiber	6.2 g	Potassium	639 mg		
Ash	1.6 g	Ascorbic.	76 mg		

No. 17 Chamomile Flowers

Amount of nutrient per 100g. (About ½ cup)				Calories: 56	
Water	81.2 g	Calcium	126 mg	β Carotene	41 µg
Protein	2.2 g	Phos.	60 mg	Thiamin	15 µg
Fat	0.7 g	Iron	3.2 mg	Riboflavin	80 µg
Carbos.	14.7 g	Sodium	48.5 mg	Niacin	2801 µg
Fiber	1.4 g	Potassium	248 mg		
Ash	1.2 g	Ascorbic.	5 mg		

No. 18 Chickweed Leaves

Amount of nutrient per 100g. (About ½ cup)				Calories: 28	
Water	91.7 g	Calcium	160 mg	β Carotene	200 µg
Protein	1.2 g	Phos.	49 mg	Thiamin	Tr.
Fat	0.2 g	Iron	29 mg	Riboflavin	100 µg
Carbos.	5.3 g	Sodium	82 mg	Niacin	500 µg
Fiber	1.7 g	Potassium	243 mg		
Ash	1.6 g	Ascorbic.	350 mg		

No. 19 Chicory

Amount of nutrient per 100g. (About ½ cup)				Calories: 30	
Water	92.0 g	Calcium	100 mg	β Carotene	2400 µg
Protein	1.7 g	Phos.	47 mg	Thiamin	100 µg
Fat	0.3 g	Iron	0.9 mg	Riboflavin	100 µg
Carbos.	4.7 g	Sodium	45 mg	Niacin	500 µg
Fiber	0.8 g	Potassium	420 mg		
Ash	1.3 g	Ascorbic.	24 mg		

No. 20 Chicory Roots

Amount of nutrient per 100g. (About ½ cup)				Calories: 73	
Water	80.0 g	Calcium	41 mg	β Carotene	Tr.
Protein	1.4 g	Phos.	61 mg	Thiamin	Tr.
Fat	0.2 g	Iron	0.8 mg	Riboflavin	Tr.
Carbos.	17.5 g	Sodium	50 mg	Niacin	400 µg
Fiber	2.0 g	Potassium	290 mg		
Ash	0.9 g	Ascorbic.	5 mg		

No. 21 Cholla Shoots

Amount of nutrient per 100g. (About ½ cup)				Calories: 31.0	
Water	77.8 g	Calcium	3 mg	β Carotene	Tr.
Protein	1.6 g	Phos.	1 mg	Thiamin	Tr.
Fat	0.5 g	Iron	1.1 mg	Riboflavin	Tr.
Carbos.	15.9 g	Sodium	0.4 mg	Niacin	Tr.
Fiber	1.7 g	Potassium	12 mg		
Ash	4.2 g	Ascorbic.	57 mg		

No. 22 Clover Flowers

Amount of nutrient per 100g. (About ½ cup)				Calories: 54	
Water	82.3 g	Calcium	232 mg	β Carotene	200 µg
Protein	2.0 g	Phos.	57 mg	Thiamin	100 µg
Fat	0.6 g	Iron	1 mg	Riboflavin	100 µg
Carbos.	13.5 g	Sodium	3 mg	Niacin	2200 µg
Fiber	1.8 g	Potassium	354 mg		
Ash	1.5 g	Ascorbic.	52 mg		

No. 23 Clover Shoots

Amount of nutrient per 100g. (About ½ cup)				Calories: 31	
Water	89.9 g	Calcium	142 mg	β Carotene	1166 µg
Protein	2.1 g	Phos.	32 mg	Thiamin	150 µg
Fat	0.5 g	Iron	1.1 mg	Riboflavin	190 µg
Carbos.	6.3 g	Sodium	4 mg	Niacin	1000 µg
Fiber	2.2 g	Potassium	345 mg		
Ash	1.1 g	Ascorbic.	57 mg		

No. 24 Daisy Flowers

Amount of nutrient per 100g. (About ½ cup)				Calories: 17	
Water	92.6 g	Calcium	56 mg	β Carotene	8800 µg
Protein	1.6 g	Phos.	32 mg	Thiamin	30 µg
Fat	0.2 g	Iron	3.1 mg	Riboflavin	20 µg
Carbos.	4.4 g	Sodium	52 mg	Niacin	860 µg
Fiber	0.9 g	Potassium	570 mg		
Ash	1.3 g	Ascorbic.	53 mg		

No. 25 Daisy Leaves

Amount of nutrient per 100g. (About ½ cup)				Calories: 19	
Water	93.5 g	Calcium	63 mg	β Carotene	3160 µg
Protein	1.8 g	Phos.	34 mg	Thiamin	90 µg
Fat	0.3 g	Iron	2.5 mg	Riboflavin	190 µg
Carbos.	3.3 g	Sodium	106 mg	Niacin	600 µg
Fiber	0.9 g	Potassium	256 mg		
Ash	1.1 g	Ascorbic.	27 mg		

No. 26 Dandelion Leaves

Amount of nutrient per 100g. (About ½ cup)				Calories: 45	
Water	85.6 g	Calcium	187 mg	β Carotene	800 µg
Protein	2.7 g	Phos.	66 mg	Thiamin	200 µg
Fat	0.7 g	Iron	3 mg	Riboflavin	300 µg
Carbos.	9.2 g	Sodium	76 mg	Niacin	900 µg
Fiber	1.6 g	Potassium	398 mg		
Ash	1.8 g	Ascorbic.	68 mg		

No. 27 Dock

Amount of nutrient per 100g. (About ½ cup)				Calories: 21	
Water	92.6 g	Calcium	74 mg	β Carotene	2770 µg
Protein	1.5 g	Phos.	56 mg	Thiamin	60 µg
Fat	0.3 g	Iron	5.6 mg	Riboflavin	80 µg
Carbos.	4.1 g	Sodium	5 mg	Niacin	400 µg
Fiber	0.9 g	Potassium	338 mg		
Ash	1.5 g	Ascorbic.	30 mg		

No. 28 Dock Seeds

Amount of nutrient per 100g. (About ½ cup)				Calories: 342	
Water	14.3 g	Calcium	35 mg	β Carotene	0 µg
Protein	12.9 g	Phos.	305 mg	Thiamin	450 µg
Fat	3.8 g	Iron	3.4 mg	Riboflavin	200 µg
Carbos.	66.9 g	Sodium	4 mg	Niacin	3800 µg
Fiber	19.3 g	Potassium	339 mg		
Ash	2.8 g	Ascorbic.	0 mg		

No. 29 Evening Primrose Seeds

Amount of nutrient per 100g. (About ½ cup)				Calories: 342	
Water	7.0 g	Calcium	1422 mg	β Carotene	0 µg
Protein	15.1 g	Phos.	533 mg	Thiamin	450 µg
Fat	2.2 g	Iron	22 mg	Riboflavin	200 µg
Carbos.	50.4 g	Sodium	16 mg	Niacin	3800 µg
Fiber	3.6 g	Potassium	542 mg		
Ash	5.3 g	Ascorbic.	0 mg		

No. 30 Filarie Shoots

Amount of nutrient per 100g. (About ½ cup)				Calories: 31	
Water	81.5 g	Calcium	464 mg	β Carotene	1166 µg
Protein	2.4 g	Phos.	87 mg	Thiamin	150 µg
Fat	0.4 g	Iron	1.1 mg	Riboflavin	190 µg
Carbos.	12.2 g	Sodium	4 mg	Niacin	1000 µg
Fiber	4.5 g	Potassium	622 mg		
Ash	3.5 g	Ascorbic.	57 mg		

No. 31 Filarie Leaves

Amount of nutrient per 100g. (About ½ cup)				Calories: 32	
Water	85.8 g	Calcium	140 mg	β Carotene	4000 µg
Protein	4.0 g	Phos.	50 mg	Thiamin	100 µg
Fat	0.4 g	Iron	2.5 mg	Riboflavin	200 µg
Carbos.	7.3 g	Sodium	60 mg	Niacin	700 µg
Fiber	1.4 g	Potassium	410 mg		
Ash	2.5 g	Ascorbic.	63 mg		

No. 32 Fireweed Greens

Amount of nutrient per 100g. (About ½ cup)				Calories: 35	
Water	88.0 g	Calcium	186 mg	β Carotene	5652 µg
Protein	2.8 g	Phos.	57 mg	Thiamin	110 µg
Fat	0.7 g	Iron	2.7 mg	Riboflavin	240 µg
Carbos.	8.2 g	Sodium	40 mg	Niacin	900 µg
Fiber	1.3 g	Potassium	382 mg		
Ash	0.8 g	Ascorbic.	68 mg		

No. 33 Goldenrod Shoots

Amount of nutrient per 100g. (About ½ cup)				Calories: 31	
Water	89.9 g	Calcium	113 mg	β Carotene	300 µg
Protein	0.9 g	Phos.	52 mg	Thiamin	150 µg
Fat	1.4 g	Iron	11 mg	Riboflavin	190 µg
Carbos.	7.0 g	Sodium	4 mg	Niacin	1000 µg
Fiber	1.3 g	Potassium	83 mg		
Ash	0.7 g	Ascorbic.	1000 mg		

No. 34 Grape Fruit

Amount of nutrient per 100g. (About ½ cup)				Calories: 63	
Water	81.3 g	Calcium	14 mg	β Carotene	Tr.
Protein	0.6 g	Phos.	10 mg	Thiamin	100 µg
Fat	0.4 g	Iron	0.3 mg	Riboflavin	100 µg
Carbos.	17.1 g	Sodium	2 mg	Niacin	300 µg
Fiber	0.8 g	Potassium	191 mg		
Ash	0.6 g	Ascorbic.	4 mg		

No. 35 Grape Leaves

Amount of nutrient per 100g. (About ½ cup)				Calories: 68	
Water	79.2 g	Calcium	716 mg	β Carotene	5200 µg
Protein	4.8 g	Phos.	38 mg	Thiamin	200 µg
Fat	0.9 g	Iron	7 mg	Riboflavin	100 µg
Carbos.	13.5 g	Sodium	21 mg	Niacin	1200 µg
Fiber	2.1 g	Potassium	382 mg		
Ash	1.5 g	Ascorbic.	34 mg		

No. 36 Lamb's Quarters Seeds

Amount of nutrient per 100g. (About ½ cup)				Calories: 353	
Water	8.3 g	Calcium	1036 mg	β Carotene	0 µg
Protein	19.6 g	Phos.	340 mg	Thiamin	450 µg
Fat	4.2 g	Iron	64 mg	Riboflavin	200 µg
Carbos.	57.7 g	Sodium	9 mg	Niacin	3800 µg
Fiber	27.1 g	Potassium	1687 mg		
Ash	10.2 g	Ascorbic.	0 mg		

No. 37 Lamb's Quarters Shoots

Amount of nutrient per 100g. (About ½ cup)				Calories: 35	
Water	88.0 g	Calcium	324 mg	β Carotene	3800 μg
Protein	3.5 g	Phos.	48 mg	Thiamin	100 μg
Fat	0.8 g	Iron	1.5 mg	Riboflavin	200 μg
Carbos.	5.5 g	Sodium	4 mg	Niacin	1000 μg
Fiber	2.0 g	Potassium	684 mg		
Ash	2.4 g	Ascorbic.	40 mg		

No. 38 Mallow (Malva) Leaves

Amount of nutrient per 100g. (About ½ cup)				Calories: 28	
Water	91.6 g	Calcium	90 mg	β Carotene	3315 μg
Protein	3.6 g	Phos.	42 mg	Thiamin	170 μg
Fat	1.4 g	Iron	3.7 mg	Riboflavin	290 μg
Carbos.	2.1 g	Sodium	60 mg	Niacin	500 μg
Fiber	0.9 g	Potassium	410 mg		
Ash	1.3 g	Ascorbic.	24 mg		

No. 39 Maple Leaves

Amount of nutrient per 100g. (About ½ cup)				Calories: 1	
Water	88.0 g	Calcium	186 mg	β Carotene	5652 μg
Protein	3.3 g	Phos.	57 mg	Thiamin	110 μg
Fat	1.0 g	Iron	2.7 mg	Riboflavin	240 μg
Carbos.	6.7 g	Sodium	40 mg	Niacin	900 μg
Fiber	2.0 g	Potassium	382 mg		
Ash	0.9 g	Ascorbic.	68 mg		

No. 40 Meadowsweet Leaves

Amount of nutrient per 100g. (About ½ cup)				Calories: 31	
Water	89.0 g	Calcium	140 mg	β Carotene	4000 μg
Protein	2.8 g	Phos.	50 mg	Thiamin	100 μg
Fat	0.4 g	Iron	2.5 mg	Riboflavin	200 μg
Carbos.	6.0 g	Sodium	60 mg	Niacin	700 μg
Fiber	1.2 g	Potassium	410 mg		
Ash	1.8 g	Ascorbic.	63 mg		

No. 41 Milk Thistle Seeds

Amount of nutrient per 100g. (About ½ cup)				Calories: 342	
Water	11.1 g	Calcium	35 mg	β Carotene	0 µg
Protein	16.9 g	Phos.	305 mg	Thiamin	450 µg
Fat	25.7 g	Iron	3.4 mg	Riboflavin	200 µg
Carbos.	44.7 g	Sodium	4 mg	Niacin	3800 µg
Fiber	2.6 g	Potassium	339 mg		
Ash	1.6 g	Ascorbic.	0 mg		

No. 42 Milk Thistle Shoots

Amount of nutrient per 100g. (About ½ cup)				Calories: 31	
Water	89.9 g	Calcium	65 mg	β Carotene	300 µg
Protein	1.4 g	Phos.	70 mg	Thiamin	150 µg
Fat	0.6 g	Iron	11 mg	Riboflavin	190 µg
Carbos.	7.2 g	Sodium	4 mg	Niacin	1000 µg
Fiber	0.6 g	Potassium	83 mg		
Ash	0.9 g	Ascorbic.	1000 mg		

No. 43 Milkweed Shoots

Amount of nutrient per 100g. (About ½ cup)				Calories: 31	
Water	89.9 g	Calcium	31 mg	β Carotene	1166 µg
Protein	3.5 g	Phos.	59 mg	Thiamin	150 µg
Fat	0.5 g	Iron	1.1 mg	Riboflavin	190 µg
Carbos.	5.2 g	Sodium	4 mg	Niacin	1000 µg
Fiber	0.6 g	Potassium	345 mg		
Ash	0.9 g	Ascorbic.	57 mg		

No. 44 Mint Leaves

Amount of nutrient per 100g. (About ½ cup)				Calories: 51	
Water	84.0 g	Calcium	160 mg	β Carotene	3000 µg
Protein	2.6 g	Phos.	80 mg	Thiamin	80 µg
Fat	0.8 g	Iron	13 mg	Riboflavin	160 µg
Carbos.	10.9 g	Sodium	40 mg	Niacin	650 µg
Fiber	1.9 g	Potassium	360 mg		
Ash	1.7 g	Ascorbic.	24 mg		

No. 45 Mullein Leaves

Amount of nutrient per 100g. (About ½ cup)				Calories: 62	
Water	78.6 g	Calcium	285 mg	β Carotene	900 μg
Protein	2.3 g	Phos.	122 mg	Thiamin	Tr.
Fat	0.3 g	Iron	50 mg	Riboflavin	Tr.
Carbos.	17.1 g	Sodium	16 mg	Niacin	2000 μg
Fiber	2.4 g	Potassium	282 mg		
Ash	1.8 g	Ascorbic.	16 mg		

No. 46 Mustard Leaves

Amount of nutrient per 100g. (About ½ cup)				Calories: 31	
Water	89.5 g	Calcium	183 mg	β Carotene	4200 μg
Protein	3.0 g	Phos.	50 mg	Thiamin	110 μg
Fat	0.5 g	Iron	3 mg	Riboflavin	220 μg
Carbos.	5.6 g	Sodium	32 mg	Niacin	800 μg
Fiber	1.1 g	Potassium	377 mg		
Ash	1.4 g	Ascorbic.	97 mg		

No. 47 Mustard Seeds

Amount of nutrient per 100g. (About ½ cup)				Calories: 416	
Water	7.5 g	Calcium	320 mg	β Carotene	0.0 μg
Protein	22.2 g	Phos.	650 mg	Thiamin	540 μg
Fat	18.3 g	Iron	3.4 mg	Riboflavin	400 μg
Carbos.	45.6 g	Sodium	4 mg	Niacin	3800 μg
Fiber	15.9 g	Potassium	339 mg		
Ash	6.4 g	Ascorbic.	0 mg		

No. 48 Nettle Leaves

Amount of nutrient per 100g. (About ½ cup)				Calories: 35	
Water	88.2 g	Calcium	34 mg	β Carotene	1100 μg
Protein	1.2 g	Phos.	5 mg	Thiamin	100 μg
Fat	2.7 g	Iron	5 mg	Riboflavin	100 μg
Carbos.	6.9 g	Sodium	1 mg	Niacin	600 μg
Fiber	1.3 g	Potassium	20 mg		
Ash	1.0 g	Ascorbic.	10 mg		

No. 49 Phragmities Shoots

Amount of nutrient per 100g. (About ½ cup)				Calories: 31	
Water	89.9 g	Calcium	48 mg	β Carotene	1166 µg
Protein	0.5 g	Phos.	6 mg	Thiamin	150 µg
Fat	0.1 g	Iron	1.1 mg	Riboflavin	190 µg
Carbos.	9.0 g	Sodium	4 mg	Niacin	1000 µg
Fiber	4.1 g	Potassium	345 mg		
Ash	0.4 g	Ascorbic.	57 mg		

No. 50 Pine Nuts

Amount of nutrient per 100g. (About ½ cup)				Calories: 634	
Water	3.7 g	Calcium	14 mg	β Carotene	20 µg
Protein	15.3 g	Phos.	515 mg	Thiamin	760 µg
Fat	61.3 g	Iron	4 mg	Riboflavin	230 µg
Carbos.	16.8 g	Sodium	72 mg	Niacin	9800 µg
Fiber	2.6 g	Potassium	628 mg		
Ash	2.9 g	Ascorbic.	1 mg		

No. 51 Pine Needles

Amount of nutrient per 100g. (About ½ cup)				Calories: 141	
Water	51.5 g	Calcium	186 mg	β Carotene	5652 µg
Protein	3.1 g	Phos.	57 mg	Thiamin	110 µg
Fat	4.5 g	Iron	2.7 mg	Riboflavin	240 µg
Carbos.	39.9 g	Sodium	40 mg	Niacin	900 µg
Fiber	14.1 g	Potassium	382 mg		
Ash	1.0 g	Ascorbic.	68 mg		

No. 52 Plantain Leaves

Amount of nutrient per 100g. (About ½ cup)				Calories: 61	
Water	81.4 g	Calcium	184 mg	β Carotene	2520 µg
Protein	2.5 g	Phos.	52 mg	Thiamin	95 µg
Fat	0.3 g	Iron	1.2 mg	Riboflavin	280 µg
Carbos.	14.6 g	Sodium	16 mg	Niacin	800 µg
Fiber	2.0 g	Potassium	277 mg		
Ash	1.2 g	Ascorbic.	8 mg		

No. 53 Plantain Seeds

Amount of nutrient per 100g. (About ½ cup)				Calories: 342	
Water	11.2 g	Calcium	35 mg	β Carotene	0 µg
Protein	17.0 g	Phos.	305 mg	Thiamin	450 µg
Fat	7.6 g	Iron	3.4 mg	Riboflavin	200 µg
Carbos.	59.7 g	Sodium	4 mg	Niacin	3800 µg
Fiber	13.7 g	Potassium	339 mg		
Ash	4.5 g	Ascorbic.	0 mg		

No. 54 Prickly Pear Fruit

Amount of nutrient per 100g. (About ½ cup)				Calories: 67	
Water	81.4 g	Calcium	57 mg	β Carotene	100 µg
Protein	1.1 g	Phos.	32 mg	Thiamin	10 µg
Fat	0.4 g	Iron	1.2 mg	Riboflavin	20 µg
Carbos.	16.6 g	Sodium	1 mg	Niacin	300 µg
Fiber	1.1 g	Potassium	178 mg		
Ash	0.5 g	Ascorbic.	18 mg		

No. 55 Purslane Shoots

Amount of nutrient per 100g. (About ½ cup)				Calories: 16	
Water	93.9 g	Calcium	65 mg	β Carotene	100 µg
Protein	1.3 g	Phos.	44 mg	Thiamin	Tr.
Fat	0.1 g	Iron	2 mg	Riboflavin	100 µg
Carbos.	3.4 g	Sodium	45 mg	Niacin	500 µg
Fiber	0.8 g	Potassium	494 mg		
Ash	1.2 g	Ascorbic.	21 mg		

No. 56 Queen Anne's Lace Roots

Amount of nutrient per 100g. (About ½ cup)				Calories: 37	
Water	89.6 g	Calcium	36 mg	β Carotene	7000 µg
Protein	1.1 g	Phos.	38 mg	Thiamin	60 µg
Fat	0.3 g	Iron	1.2 mg	Riboflavin	50 µg
Carbos.	8.3 g	Sodium	70 mg	Niacin	700 µg
Fiber	0.9 g	Potassium	245 mg		
Ash	0.7 g	Ascorbic.	8 mg		

No. 57 Raspberry Fruit

Amount of nutrient per 100g. (About ½ cup)				Calories: 73	
Water	80.8 g	Calcium	30 mg	β Carotene	0 µg
Protein	1.5 g	Phos.	22 mg	Thiamin	30 µg
Fat	1.4 g	Iron	0.9 mg	Riboflavin	90 µg
Carbos.	15.7 g	Sodium	1 mg	Niacin	900 µg
Fiber	5.1 g	Potassium	199 mg		
Ash	0.6 g	Ascorbic.	18 mg		

No. 58 Raspberry Leaves

Amount of nutrient per 100g. (About ½ cup)				Calories: 46	
Water	83.1 g	Calcium	204 mg	β Carotene	1900 µg
Protein	1.9 g	Phos.	40 mg	Thiamin	100 µg
Fat	0.3 g	Iron	17 mg	Riboflavin	Tr.
Carbos.	13.4 g	Sodium	1 mg	Niacin	6500 µg
Fiber	1.4 g	Potassium	3226 mg		
Ash	1.4 g	Ascorbic.	62 mg		

No. 59 Rose Flowers

Amount of nutrient per 100g. (About ½ cup)				Calories: 16	
Water	94.8 g	Calcium	47 mg	β Carotene	400 µg
Protein	1.4 g	Phos.	86 mg	Thiamin	20 µg
Fat	0.3 g	Iron	1.0 mg	Riboflavin	110 µg
Carbos.	2.7 g	Sodium	18 mg	Niacin	600 µg
Fiber	0.6 g	Potassium	166 mg		
Ash	0.8 g	Ascorbic.	18 mg		

No. 60 Rose Hips

Amount of nutrient per 100g. (About ½ cup)				Calories: 54	
Water	91.3 g	Calcium	19 mg	β Carotene	300 µg
Protein	0.8 g	Phos.	18 mg	Thiamin	Tr.
Fat	0.4 g	Iron	0.6 mg	Riboflavin	Tr.
Carbos.	7.0 g	Sodium	1.2 mg	Niacin	400 µg
Fiber	1.1 g	Potassium	178 mg		
Ash	0.5 g	Ascorbic.	23 mg		

No. 61 Saguaro

Amount of nutrient per 100g. (About ½ cup)				Calories: 122	
Water	37.5 g	Calcium	43 mg	β Carotene	634 µg
Protein	13.1 g	Phos.	41 mg	Thiamin	95 µg
Fat	1.0 g	Iron	1.4 mg	Riboflavin	97 µg
Carbos.	41.5 g	Sodium	2 mg	Niacin	906 µg
Fiber	13.4 g	Potassium	403 mg		
Ash	6.9 g	Ascorbic.	52 mg		

No. 62 Saguaro Seeds

Amount of nutrient per 100g. (About ½ cup)				Calories: 528	
Water	4.8 g	Calcium	95 mg	β Carotene	0 µg
Protein	15.5 g	Phos.	305 mg	Thiamin	450 µg
Fat	29.1 g	Iron	3.4 mg	Riboflavin	200 µg
Carbos.	47.7 g	Sodium	4 mg	Niacin	3800 µg
Fiber	2.6 g	Potassium	339 mg		
Ash	2.9 g	Ascorbic.	0 mg		

No. 63 Sheep Sorrel (Wood Sorrel) Leaves

Amount of nutrient per 100g. (About ½ cup)				Calories: 28	
Water	90.0 g	Calcium	66 mg	β Carotene	7740 µg
Protein	2.1 g	Phos.	41 mg	Thiamin	75 µg
Fat	0.3 g	Iron	1.6 mg	Riboflavin	100 µg
Carbos.	5.6 g	Sodium	5 mg	Niacin	465 µg
Fiber	0.8 g	Potassium	338 mg		
Ash	1.1 g	Ascorbic.	119 mg		

No. 64 Shepherd's Purse Leaves

Amount of nutrient per 100g. (About ½ cup)				Calories: 33	
Water	88.2 g	Calcium	208 mg	β Carotene	2590 µg
Protein	4.2 g	Phos.	86 mg	Thiamin	250 µg
Fat	0.5 g	Iron	4.8 mg	Riboflavin	170 µg
Carbos.	5.2 g	Sodium	40 mg	Niacin	400 µg
Fiber	1.2 g	Potassium	394 mg		
Ash	1.9 g	Ascorbic.	36 mg		

No. 65 Sow Thistle Leaves

Amount of nutrient per 100g. (About ½ cup)				Calories: 31	
Water	31 g	Calcium	146 mg	β Carotene	4500 µg
Protein	0.5 g	Phos.	44 mg	Thiamin	100 µg
Fat	0.8 g	Iron	2 mg	Riboflavin	200 µg
Carbos.	6.6 g	Sodium	32 mg	Niacin	700 µg
Fiber	2.0 g	Potassium	299 mg		
Ash	1.5 g	Ascorbic.	13 mg		

No. 66 Sow Thistle Shoots

Amount of nutrient per 100g. (About ½ cup)				Calories: 20	
Water	92.0 g	Calcium	93 mg	β Carotene	116 µg
Protein	2.4 g	Phos.	35 mg	Thiamin	150 µg
Fat	0.5 g	Iron	8 mg	Riboflavin	190 µg
Carbos.	4.2 g	Sodium	4 mg	Niacin	1000 µg
Fiber	0.6 g	Potassium	67 mg		
Ash	0.9 g	Ascorbic.	32 mg		

No. 67 Strawberry Fruit

Amount of nutrient per 100g. (About ½ cup)				Calories: 37	
Water	89.9 g	Calcium	21 mg	β Carotene	36 µg
Protein	0.7 g	Phos.	21 mg	Thiamin	30 µg
Fat	0.5 g	Iron	1 mg	Riboflavin	70 µg
Carbos.	8.4 g	Sodium	1 mg	Niacin	600 µg
Fiber	1.3 g	Potassium	164 mg		
Ash	0.5 g	Ascorbic.	59 mg		

No. 68 Sumac Fruit

Amount of nutrient per 100g. (About ½ cup)				Calories: 230	
Water	4.8 g	Calcium	93 mg	β Carotene	1700 µg
Protein	11.9 g	Phos.	57 mg	Thiamin	300 µg
Fat	16.8 g	Iron	10 mg	Riboflavin	300 µg
Carbos.	63.2 g	Sodium	95 mg	Niacin	2500 µg
Fiber	27.8 g	Potassium	980 mg		
Ash	3.2 g	Ascorbic.	143 mg		

No. 69 Sunflower Seeds

Amount of nutrient per 100g. (About ½ cup)				Calories: 560	
Water	4.8 g	Calcium	120 mg	β Carotene	30 μg
Protein	24.0 g	Phos.	837 mg	Thiamin	1960 μg
Fat	47.3 g	Iron	7 mg	Riboflavin	230 μg
Carbos.	19.9 g	Sodium	30 mg	Niacin	5400 μg
Fiber	3.8 g	Potassium	920 mg		
Ash	4.0 g	Ascorbic.	0 mg		

No. 70 Thistle Leaves

Amount of nutrient per 100g. (About ½ cup)				Calories: 63	
Water	80.9 g	Calcium	20 mg	β Carotene	0 μg
Protein	3.0 g	Phos.	79 mg	Thiamin	Tr.
Fat	0.1 g	Iron	8.5 mg	Riboflavin	Tr.
Carbos.	15.3 g	Sodium	65 mg	Niacin	200 μg
Fiber	1.5 g	Potassium	410 mg		
Ash	0.7 g	Ascorbic.	0 mg		

No. 71 Thyme Leaves

Amount of nutrient per 100g. (About ½ cup)				Calories: 60	
Water	80.0 g	Calcium	27 mg	β Carotene	400 μg
Protein	2.0 g	Phos.	44 mg	Thiamin	100 μg
Fat	1.6 g	Iron	27 mg	Riboflavin	100 μg
Carbos.	13.9 g	Sodium	12 mg	Niacin	1100 μg
Fiber	0.4 g	Potassium	177 mg		
Ash	2.6 g	Ascorbic.	113 mg		

No. 72 Tumbleweed Seeds

Amount of nutrient per 100g. (About ½ cup)				Calories: 369	
Water	4.2 g	Calcium	830 mg	β Carotene	0 μg
Protein	27.4 g	Phos.	540 mg	Thiamin	500 μg
Fat	14.4 g	Iron	143 mg	Riboflavin	200 μg
Carbos.	46.3 g	Sodium	11 mg	Niacin	4100 μg
Fiber	30.2 g	Potassium	1140 mg		
Ash	7.7 g	Ascorbic.	0 mg		

No. 73 Tumbleweed Shoots

Amount of nutrient per 100g. (About ½ cup)				Calories: 113	
Water	60.6 g	Calcium	973 mg	β Carotene	3500 µg
Protein	4.8 g	Phos.	67 mg	Thiamin	100 µg
Fat	0.7 g	Iron	5 mg	Riboflavin	200 µg
Carbos.	27.9 g	Sodium	4 mg	Niacin	1000 µg
Fiber	12.5 g	Potassium	2545 mg		
Ash	6.0 g	Ascorbic.	40 mg		

No. 74 Violet Flowers

Amount of nutrient per 100g. (About ½ cup)				Calories: 16	
Water	94.8 g	Calcium	47 mg	β Carotene	400 µg
Protein	1.4 g	Phos.	86 mg	Thiamin	20 µg
Fat	0.3 g	Iron	1 mg	Riboflavin	110 µg
Carbos.	2.7 g	Sodium	18 mg	Niacin	600 µg
Fiber	0.6 g	Potassium	166 mg		
Ash	0.8 g	Ascorbic.	18 mg		

No. 75 Wild Lettuce Leaves

Amount of nutrient per 100g. (About ½ cup)				Calories: 18	
Water	94.0 g	Calcium	68 mg	β Carotene	1000 µg
Protein	1.3 g	Phos.	25 mg	Thiamin	60 µg
Fat	0.3 g	Iron	1.4 mg	Riboflavin	60 µg
Carbos.	3.5 g	Sodium	9 mg	Niacin	400 µg
Fiber	0.7 g	Potassium	264 mg		
Ash	0.9 g	Ascorbic.	18 mg		

No. 76 Willow Leaves

Amount of nutrient per 100g. (About ½ cup)				Calories: 35	
Water	90.0 g	Calcium	92 mg	β Carotene	5652 µg
Protein	1.6 g	Phos.	25 mg	Thiamin	110 µg
Fat	0.3 g	Iron	2.7 mg	Riboflavin	240 µg
Carbos.	7.3 g	Sodium	40 mg	Niacin	900 µg
Fiber	1.6 g	Potassium	382 mg		
Ash	0.8 g	Ascorbic.	68 mg		

No. 77 Wintergreen Fruit

Amount of nutrient per 100g. (About ½ cup)				Calories: 68	
Water	81.7 g	Calcium	69 mg	β Carotene	280 µg
Protein	2.4 g	Phos.	18 mg	Thiamin	42 µg
Fat	0.9 g	Iron	0.6 mg	Riboflavin	43 µg
Carbos.	14.4 g	Sodium	1 mg	Niacin	400 µg
Fiber	1.3 g	Potassium	178 mg		
Ash	0.5 g	Ascorbic.	89 mg		

No. 78 Yarrow Leaves

Amount of nutrient per 100g. (About ½ cup)				Calories: 103	
Water	65.4 g	Calcium	460 mg	β Carotene	Tr.
Protein	4.5 g	Phos.	125 mg	Thiamin	Tr.
Fat	1.2 g	Iron	Tr.	Riboflavin	Tr.
Carbos.	25.2 g	Sodium	3 mg	Niacin	Tr.
Fiber	5.9 g	Potassium	616 mg		
Ash	3.7 g	Ascorbic.	23 mg		

The first "A Survival Acre" at Linda's home wild food tours.

Burke's Cottages Sabael, NY wild food walk.

Runyon Institute, Warrensburg, NY wild food walk.

Part III: Wild Food Recipes

These recipes incorporate wild foods as much as possible. As you will see, wild foods can be used in all types of dishes including those that use meat, fish, and dairy foods. Of course, dishes made entirely of wild foods are *my* environmentarian preference. Those included here use meats, fish, dairy foods, and other "nonwild" foods and are adapted from various cultures.

The important thing is to *try* wild foods. Fit them into your diet occasionally at first, increasing to a daily basis. Supplementing your meals with wild food saves money. Most of the recipes may seem unfamiliar, but dare to use wild foods in a creative way and you will be pleasantly surprised!

A typical campfire meal: Food simmering on edge, drying over hot coals, casserole on back, hot cookies or biscuits on back rock.

Some suggestions for substitutions:

1. Anise flavor: Goldenrod leaves, flowers, and roots.
2. Asparagus: Milkweed shoots, fireweed tops.
3. Bran: Bulrush, crabgrass.
4. Broccoli: Milkweed buds and flowers, wintercress, mustard flowers.
5. Cabbage: Shepherd's purse, dandelion, chicory, and wild lettuce.
6. Carrot: Queen Anne's lace roots, evening primrose roots, thistle roots, Shepherd's purse roots, and purslane roots.
7. Coffee: Chicory, dandelion roots, yarrow, clover buds.
8. Corn on the cob: Green cattail heads.
9. Cucumbers: cattail pith, fireweed pith, sunflower pith.
10. Endive: Dock.
11. Fruit: Blackberries, blueberries, dandelion flowers, mullein flowers, evening primrose flowers, raspberries, roses, strawberries, and violets.
12. Granola: Phragmities, amaranth, lamb's quarters, clover.
13. Gum: Sap from balsam fir or pine, milk of wild lettuce, dandelion sap, milkweed sap.
14. Lemonade: Sumac, wood sorrel, and sheep sorrel.
15. Lettuce: Sow thistle, dandelion, chicory, wild lettuce, milk thistle, amaranth, malva neglecta, and lamb's quarters leaves.
16. Nuts: Pine seeds, seeds from malva neglecta.
17. Okra: Milkweed shoots, flowers, pods, small leaves.
18. Orange juice: Pine tea, sumac tea, and balsam fir tea.
19. Pepper: Thyme leaves and flowers.
20. Potatoes: Arrowhead, ground pine nuts.
21. Salt: Queen Anne's lace seeds, ground dried clover with dried garlic.
22. Sugar: Dried meadowsweet flowers.
23. Turnip greens: Plantain leaves, sow thistle leaves.

Some medicinal and cosmetic substitutes.

1. Aspirin: Willow.
2. Lampwick: Mullein leaf, dried and cut in strips.
3. Mothballs: Dried white clover.
4. Sedative: Chamomile, mullein.

Some Wild Food Menus

During the Runyon Institute days, students from Rutgers University experimented with recipes. Some recipes are from homestead days while others are shared with us by students of wild food classes. Still others are delectable recipes from gourmet cooks. To help clarify which recipes fit these groups please see the following symbols listed below.

Food Group Symbols
* * From Institute students
* ! Homestead days
* x Given to us by Arizona students and individual people over the last ten years.
* ~ Gourmet cooks

Because of my vegetarian habits, the following recipes do not use meat or fish. Wild food recipes can be combined in a variety of ways and still fit the traditional meals of breakfast, lunch and dinner. Here are some menu suggestions to get you started on the road to an environmentarian diet.

Soups & Salads

Burdock Stem Soup*

    2 cups 1-inch sections of burdock stems, peeled and sliced
    1 quart water
    2 tablespoons lemon juice or vinegar
    pinch of thyme pepper, Queen Anne's lace seeds to taste

Place burdock stems in a medium pot and cover with 2 cups water. Bring to a boil and simmer for 5 minutes. Drain. Add remaining 2 cups water, thyme pepper, and seeds. Bring to a boil and simmer for 5 minutes. Season with lemon juice or vinegar and serve hot.

Serves 4

Slow Cooker Filarie Soup x

    1 large filarie head and root
    1 medium onion, thinly sliced
    1 small potato, peeled and diced (can substitute parsnips or Queen Anne's lace root)

Wash the filarie head well, pulling out the buds and fruit. Scrub the root with a toothbrush. Place the filarie in a slow cooker and cover with water. Add the onion and potato, cover with water and simmer on high for 1 ½ hours. Serve hot.

Serves 4

Clover Soup !

    1 1/2 cups chopped onions
    6 cups water
    3 cups clover leaves and flowers washed
    4 teaspoons tamari or soy sauce
    1 teaspoon wild thyme
    salt or Queen Anne's lace seeds to taste
    3 cups cooked brown rice (optional)

Simmer the onions in the water for 20 minutes. Add the remaining ingredients, except rice, and simmer 10 minutes. Eat as broth or stir in the brown rice and cook until the rice is done.

Variation: Add 2 cups chopped dandelion root and 1 cup sliced carrots to the water and simmer with the onion.

Serves 6

Lamb's Quarters Fat Hen Soup x

    1 pound lamb's quarters leaves and seeds (6 to 8 cups), washed
    1 large onion, chopped
    1 teaspoon salt or Queen Anne's lace seeds
    1/4 teaspoon freshly ground black or thyme pepper
    1/4 teaspoon freshly grated nutmeg
    1 garlic clove, crushed
    2 tablespoons butter or olive oil
    2 teaspoons whole wheat flour
    1/2 cup milk or water
    1/2 cup sour cream or water
    2 teaspoons pine nuts

Boil the lamb's quarters in water to cover until tender, about 5 minutes. Add onion, salt or seeds, pepper, and nutmeg. Cover and simmer 10 minutes. Remove from the heat and cool.

Put the mixture in a blender and puree to a smooth, deep green. Add the garlic and blend again. Melt the butter in a medium saucepan over moderate heat. Stir in the flour and cook 1/2 minute. Add the milk, then sour cream or water to make a white sauce. Stir to blend, then serve lamb's quarters into bowls, top with white sauce and garnish with pine nuts.

Serves 6

Delicate Mallow Soup!

    1 cup malva neglecta leaves, washed
    2 cups water
    1 small onion, thinly sliced
    1/4 cup malva neglecta seeds

Combine the leaves and water in a medium pot. Bring to a boil over medium heat, then turn off heat, stir and cover. Cool for 10 minutes.

Strain the leaves if you wish, then add the onion and seeds. Serve hot.

Serves 2

Simple Nettle Soup!

> 3 cups nettles, washed and sliced (use any part of whole plant: leaves, flowers, seeds, or stems)
> 4 cups water
> 2 tablespoons butter
> 3 garlic cloves, crushed
> salt or Queen Anne's lace seeds to taste

Combine all ingredients in a large pot and simmer for 20 minutes. Serve hot with whole-grain bread.

Serves 4

Nettle Chowder x

> 4 cups water
> 1 1/2 cups diced potatoes
> 1 cup sliced carrots (wild if available)
> 2 celery stalks, sliced
> 1 cup chopped onion
> 3 garlic cloves, crushed
> 1/2 teaspoon fresh or dried marjoram
> 1 teaspoon dried or fresh thyme leaves
> Salt or Queen Anne's lace seeds to taste
> 4 cups nettles, washed and sliced (any part of whole plant: leaves, flowers, seeds, or stems)
> 2 cups soymilk

Combine the water, potatoes, carrots, celery, onion, herbs, and seeds in a large pot. Simmer 30 minutes, then add nettles and simmer another 20 minutes. Add soymilk and heat gently, but do not boil. Serve hot!

Serves 6

Sesame Thistle Soup*

>3 tablespoons butter
>1 cup chopped thistle root (any species)
>3 cups water
>2/3 cup sesame seeds
>4 garlic cloves, crushed
>1 cup dandelion flowers
>Queen Anne's lace seeds or salt to taste
>3 cups cooked brown rice

Melt the butter in a medium skillet. Sauté the thistle, sesame seeds, and garlic until the seeds are golden brown, about 2 minutes. Add the dandelion and cook another 5 minutes. Add the seeds or salt to taste, then serve over brown rice.
>**Serves 6**

Milk Thistle Soup with Yellow Squash x

>6-8 milk thistle leaves
>1 cup cooked, pureed yellow squash
>garlic bud (optional)

Place milk thistle leaves in water and simmer until soft and pliable (5-10 minutes). Puree in a blender. Place milk thistle puree and squash puree in two separate pots and heat. Pour in shallow soup dish, one side at a time so you have a dark half and a light half (ying and yang) in your bowl. Press garlic bud and sprinkle over the top.) Serve hot!
>**Serves 2**

Sow Thistle Soup x

>1 cup sow thistle tips
>1 clove garlic
>1 onion, thinly sliced

Place sow thistle tips in water to cover. Simmer 20 minutes on low heat. (Crock pot is excellent; simmer for 1 hour on low.) Add garlic, onion, and serve hot.
>**Serves 2**

Tumbleweed Soup x

    1 cup tumbleweed tips
    1 clove garlic
    1 thinly sliced onion

Place tips in water to cover. Simmer 20 minutes on low heat. Or use a crock pot for more flavor and cook at low heat for 1 hour. Add pressed oil from garlic and garnish with onion.
Serves 2

Greens Soup *

    1 cup fresh sorrel leaves, flowers, stems
    1 thinly sliced onion

Place sorrel in a pot and cover with water. Add onion and simmer for a few minutes. Leaves will turn from bright green to olive green. This is normal and ensures the oxalic acid crystals are destroyed. Serve hot. Delicious!
Serves 2

Creamed Sorrel Soup ~

    4 cups washed sorrel leaves
    2 teaspoons safflower oil or olive oil
    2 teaspoons wheat germ
    1 onion, chopped
    4 cups non-fat dried milk, or goat's milk, or water

Place leaves in pot, cover with water and simmer slowly for 1/2 hour. Blend in remaining ingredients, simmer and strain.
Serves 4

Wild Lettuce Salad *

    1 cup wild lettuce leaves, washed
    1/2 cup wild lettuce buds and flowers, washed
    1/2 cup shredded red cabbage
    1 garlic clove, crushed
    1 teaspoon wild thyme leaves fresh or dried
    3 tablespoons oil, 3 tablespoons vinegar of choice

Combine all ingredients, toss gently, and serve.
Serves 4

Fresh Dandelion Salad !

>4 cups fresh dandelion greens
>clove of garlic, chopped, or garlic salt
>pinch of thyme pepper

Drain the dandelion greens. Rub the garlic on a salad bowl. Add chopped garlic pieces to bowl, and place drained leaves in bowl. Add your favorite dressing.
>Serves 3

Hot Dandelion Salad *

>4 handfuls fresh (young, new) dandelion leaves
>sesame oil or olive oil
>thyme pepper

Wash dandelions. If desired, soak bitter leaves in salted water for about 1 hour then wash and drain. Place the leaves in a frying pan and add the oil. Heat slowly to a moderate heat while stirring the leaves until they are wilted. Remove, serve hot!
>Serves 4

Daisy Petal Salad x

>1 quart freshly picked daisy petals

Wash petals, place in salad bowl and add your favorite dressing. Or put petals in pita sandwiches with favorite garnish.
>Serves 4-6

Spring Wild Coleslaw x

>1/2 cup dandelion
>1/2 cup plantain
>1/2 cup clover
>1/2 cup chickweed
>1/4 cup curley dock
>1/2 small peeled onion
>2 tablespoons oil and vinegar

Wash all ingredients well and drain in a sieve. Process all parts plus onion in a processor until completely "slawed". Add oil and vinegar (or your favorite dressing). Serve as a side dish.

Chickweed Salad x

    1 clove garlic
    1 quart loosely packed chickweed, washed
    2 eggs, hard-cooked
    4 tablespoons favorite dressing

Peel clove of garlic and rub wooden salad bowl. Chop chickweed finely and add to salad bowl. Peel eggs, slice thin and add to bowl. Toss together with favorite salad dressing.
Serves 4

Star Chickweed Dressing x

    2 cups star chickweed, washed
    3/4 cup plain yogurt
    3/4 cup mayonnaise
    1/2 tablespoon garlic powder or 4 cloves fresh
    1 tablespoon dried dill
    salt to taste

Combine all ingredients in a blender or food processor. Blend on medium speed, then refrigerate for 30 minutes. Serve on your favorite salad or casserole. Sandwiches also taste delicious spread with this dressing.
Makes 2 cups of dressing

Evening Primrose Olive Oil*

    1 cup evening primrose flowers, washed and dried
    approx. 1\2 cup extra virgin olive oil (sunflower, safflower, or corn oil can be substituted)

Fill a 10-ounce jar to capacity with evening primrose flowers. Add oil at room temperature or boiling to cover. (Be careful with hot oil in glass jars.) Repeatedly press down the flowers until all the air bubbles are out of the mixture. Cap and shake every day for or 1 week for sunflower, safflower, or corn oil, or 2 weeks for olive oil. Remove the flowers and discard. Refrigerate oil if you used sunflower, safflower, or corn oil. Use for salads.
Makes approximately 10 ounces

Vegetable Dishes & Light Meals

Linguine with Amaranth Cream Sauce~

>2 tablespoons butter
>4 garlic cloves, crushed
>3 cups soymilk or cow's milk
>4 tablespoons rice or wheat flour
>1 cup crushed dried young amaranth
>1 pound linguine, cooked until al dente

In a large frying pan, melt the butter and sauté the garlic over low heat for about 10 minutes. In a saucepan, mix the milk and flour to make a thin paste. Add the flour mixture and amaranth and heat until smooth and thickened, about 3 minutes. Serve over the linguine.
>**Serves 2 to 3 as a main course**

Arrowhead Stew!

>1 large onion, sliced
>1 cup washed roots: thistle, dandelion, evening primrose, or filarie
>12 to 14 arrowhead tubers, washed
>1/2 teaspoon Queen Anne's lace seeds
>1 teaspoon dried thyme leaves

Combine all ingredients in a slow cooker and add water to cover. Cook on low setting for 4 hours or until tender.
>**Serves 4 as a main course**

Clover Noodles~

>1 cup clover flour (See page 78)
>3 to 5 cups unbleached white flour
>5 eggs
>1 tablespoon olive oil
>1/2 teaspoon salt or Queen Anne's lace seeds

Put clover flour into bowl and add enough water to reconstitute to a moist consistency. Stir in eggs, oil, & salt. You should have a slimy green soup. Add white flour until you make a stiff dough. Remove and knead till it won't tear when you stretch it. Let rest 30 minutes. Cut dough ball into about seven or eight equal sized balls, and run through pasta machine. Start on setting #1 and then press through 2, 3, and 5. Add flour with each run through the machine to avoid sticking.

Dry on racks of food dryer. Let cool, store in tightly sealed glass jars. Serve cooked to al dente and topped with butter or add your favorite vegetable.

Serves 8-10

In a pasta machine, clover noodles can be made by replacing 2 tablespoons of flour with 2 tablespoons of clover flour. Adjust the liquid as necessary to get the right texture, extrude and serve.

Makes approximately 4 servings

Clover Pasta with Beans and Basil x

    1 pound clover pasta
    4 tablespoons olive oil
    4 to 6 garlic cloves
    1 cup garbanzo beans (chickpeas)
    6 tablespoons water or tomato puree
    1/2 teaspoon freshly ground black pepper or
        1/4 teaspoon ground dried thyme
    12 large basil leaves
    1 tablespoon fresh parsley, chopped
    1/4 cup grated Romano cheese (optional) or
        1 teaspoon Queen Anne's lace seeds or salt

Cook the pasta until al dente, then drain and set aside. In a food processor or blender, puree olive oil and garlic. Add garbanzos and water or tomato puree and blend. Blend in the pepper, basil, parsley, and cheese or seeds and blend quickly until mixture is a coarse paste. Toss with pasta and serve.

Serves 4 as a main course

Burdock Bur Casserole*

    2 cups young burdock burs and leaves
    1 teaspoon thyme leaves
    1 teaspoon Queen Anne's lace seeds
    1 cup nonfat dry milk or water
    4 tablespoons butter or 2 teaspoons safflower, olive oil
    1 cup whole wheat bread crumbs

Preheat the oven to 350 degrees. Wash the burs & leaves and place in a large saucepan. Add water to cover and simmer for 5 minutes. Drain.

Add the thyme and seeds, cover with water and simmer for 5 another minutes. Drain and save the water for soup stock.

Put the burs in a greased 1 1/2-quart casserole dish. Add the milk or liquid and oil. Top with bread crumbs and bake for 20 minutes.

Serves 4 as a main vegetable casserole

Cattail "Corn On The Cob" !

    2 to 3 cattail heads
    safflower, olive oil, or butter
    salt or Queen Anne's lace seeds
    thyme pepper to taste

Put the cattail in a large pot of boiling water and boil for 7 minutes. Remove and serve with oil or butter, salt or seeds, and thyme.

Serves 1 as a side dish

Note: When cutting, leave 2 inches of stem for holding the cob

Cattail Casserole!

    2 cups cattail fluff, scraped from brown buds
    1 cup bread crumbs
    1 egg, beaten (optional)
    2 to 3 teaspoons water
    Queen Anne's lace seeds or salt to taste
    1/4 teaspoon thyme leaves, fresh

Preheat the oven to 325 degrees. Combine all ingredients in a greased 1 1/2-quart casserole dish and bake for 25 minutes. Serve hot.
Serves 4 as a main dish

Chickweed Pita Pockets x

2 cups chickweed, washed
1/2 cup finely chopped tender lamb's quarters leaves, washed
1/2 cup sliced carrot
1/2 cup finely diced cucumber
1 tablespoon salad dressing of your choice
2 pita breads

Toss together the chickweed, lamb's quarters, carrot, cucumber, and dressing. Fill the pita pockets and serve.
Makes 2 sandwiches

Chickweed Mediterranean x

2 tablespoons olive oil
1 garlic clove, finely chopped
2 cups whole chickweed plant without roots, washed
1 teaspoon thyme leaves

Heat the olive oil in a medium frying pan. Sauté the garlic until browned, then add the chickweed and thyme. Stir-fry until piping hot, about 10 minutes.
Serves 3 as a side dish

Hot Clover and Rice with Butter Sauce x

4 cups cooked brown rice
4 cups washed clover leaves
1 cup milk or water
1/4 pound butter or soy butter
1 teaspoon finely minced garlic, or 2 teaspoons garlic powder

Preheat the oven to 350 degrees. Grease a 2-quart casserole dish. In the casserole, combine the rice, clover, and liquid.

Melt butter in a saucepan over low heat. Add garlic and cook for 3 minutes. Stir sauce into casserole and bake for 20 minutes until piping hot.

Serves 4 as a main course

Dandelion Greens Casserole !

2 cups dandelion leaves, washed
1 cup whole wheat bread crumbs
2 cups water

Preheat the oven to 350 degrees and grease a 1-quart casserole dish. Put the leaves in a 1-quart saucepan and add water to cover. Simmer gently for 20 minutes, then drain, reserving the liquid. Chop the leaves fine and pour into the casserole dish. Add the reserved liquid and top with the bread crumbs. Bake for 25 minutes or until brown on top.

Serves 3 to 4 as a main course

Southern Italian Dandelion Leaves ~

4 cups young 2 to 3-inch dandelion leaves, washed
1 medium potato or arrowhead tubers, peeled
2 teaspoons safflower or sesame oil
1 medium onion, cubed
1 large tomato or 1 8-ounce can whole tomatoes, chopped

Combine the dandelion leaves and potato in a medium pot and cover with water. Bring to a boil and simmer for 15 minutes. Remove the potato or arrowhead, then cube and set aside. Drain the leaves, squeezing out the water. (Save the liquid for soup stock.)

Heat the oil in a large frying pan and sauté the onion slowly until limp, about 5 minutes. Combine the dandelion leaves, potato or arrowhead, and tomato, and bring to a boil. Remove from the heat and stir.

Serves 4 to 6 as a main course

Dandelion Stir Fry ~

    3 tablespoons olive oil
    1 cup chopped onion
    4 garlic cloves, crushed
    1/4 cup chopped dandelion root, scrubbed
    2 cups fresh dandelion leaves, washed
    2 cups dandelion flowers, washed
    1 teaspoon dried thyme or 1 sprig fresh thyme
    1/2 teaspoon dried marjoram leaves or 1 fresh sprig
    2 teaspoons sesame salt
    2 tablespoons cider vinegar
    2 cups cooked brown rice

Heat half the oil in a large wok or frying pan. Add the onion, garlic, and dandelion roots. Sauté for 5 minutes or until soft. Remove from the wok and set aside in a warm place.

Heat the remaining oil in the wok and add the dandelion leaves and flowers, thyme, marjoram, salt, and vinegar. Stir-fry for 10 minutes. Stir in the onion mixture and serve over brown rice.
    Serves 6 as a main course

Dandelion Roots *

    1/2 cup washed and peeled white dandelion roots,
        chopped into French fry sized pieces
    1 cup favorite pancake butter
    2 tablespoons safflower oil
    1 chopped onion

Dip the roots in the pancake batter. French fry in hot safflower oil until brown. (A wok works perfectly.) Add the chopped onion as a garnish and serve.
    Serves 2 as a side dish

Chamomile Vegetable !

Chop a whole chamomile plant into bite-sized pieces and simmer gently for 6-8 minutes. Serve hot with favorite garnish. A delicious broccoli substitute. Servings vary depending on plant size.

Tofu Scallopine with Dandelion Wine ~

    1 pound firm tofu, weighted and drained for at least 1 hour
whole wheat flour
3 tablespoons olive oil
1 bay leaf
1/2 cup Dandelion Wine (Page 246) or marsala
10 large fresh mushrooms, sliced
1 large garlic clove, finely minced
1 teaspoon chopped chives
1 teaspoon chopped fresh parsley
juice of 1/2 lemon
1 teaspoon Worcestershire sauce
Queen Anne's lace seeds or salt to taste
Ground thyme or black pepper to taste
6 lemon wedges for garnish

Slice the tofu into 12 equal portions and coat each piece with flour. Heat the oil in a large frying pan and sauté the tofu with the bay leaf until tofu is light brown on both sides- about 6 minutes. Remove the tofu and set aside.

Add the wine and reduce the liquid by one-half. Add the mushrooms, garlic, chives, parsley, lemon juice, Worcestershire sauce, seeds or salt, and thyme or pepper. Cook for 3 minutes.

Add the tofu slices and simmer for 5 minutes, adding additional wine or water if necessary to keep mixture moist. Serve the tofu slices with pan juices and garnish with a lemon wedge and parsley

Serves 6 as a main course

Rosemary Tofu x

    3 tablespoons butter or soy butter
1 pound firm tofu, weighted, drained and cubed
4 garlic cloves, crushed
1 1/2 teaspoons dried rosemary leaves
2 cups diced young lamb's quarters or amaranth leaves, washed
3 cups cooked brown rice

Melt the butter in a large frying pan. Sauté the tofu, garlic, and rosemary for 20 minutes, stirring often. During the last 5 minutes of cooking, add the leaves and sauté until tender. Serve hot over brown rice.

Serves 6 as a main course

Evening Primrose Stew !

   5 to 6 Evening primrose roots
   1 medium potato, halved
   1 medium onion, quartered
   1 teaspoon thyme leaves dried or fresh
   1 teaspoon Queen Anne's lace seeds

Wash the roots and scrub with a toothbrush, trimming off any bruised parts. Combine the roots, potato, and onion in a slow cooker and add water to cover. Cook on low for 4 to 6 hours, then serve while hot.

Serves 2 as a main course

Dock Casserole *

   2 large bunches dock, including leaves
   4 to 6 tablespoons butter or safflower oil

Preheat oven to 350 degrees. Wash and drain the dock and steam in a pot until all parts are wilted, about 2 minutes. Discard the tougher stems and finely chop the leaves and short stems. Steam the chopped mixture for about 15 minutes more, stirring frequently. Pour the mixture into a 1-quart casserole dish, dot with butter or drizzle with oil, and bake for 20 minutes or until bubbly hot.

Serves 4-6 as a main course

Lemon Fireweed x

   2 cups fireweed tips, washed
   1 lemon wedge
   4 to 5 thyme sprigs

Steam the fireweed tips until tender, about 2 minutes. Squeeze the lemon over the tips, then sprinkle on the thyme.

Serves 4 as a side dish

Filarie Sandwich x

>1 handful filarie leaves, washed
>pinch dried thyme leaves
>1/4 teaspoon each of oil and vinegar
>1 Bermuda onion, thinly sliced into rings
>2 slices whole wheat bread

Layer the ingredients between the bread slices for a "desert sandwich" treat.
>**Makes 1 sandwich**

Fireweed Casserole *

>4 cups fireweed buds, washed
>1 medium onion, chopped, or 1 cup chopped chives
>bread crumbs

Preheat the oven to 350 degrees and lightly grease a 1 1/2-quart casserole dish. Mix the buds and onion or chives and place in the casserole dish. Add water to cover. Bake for 30 minutes or until bubbly. Top with bread crumbs.
>**Serves 4**

Wild Sow Thistle Stir-Fry !

>1 cup wild sow thistle leaves
>1/2 cup wild sow thistle buds and flowers
>1/2 cup red cabbage
>1 clove garlic, pressed
>1 teaspoon wild thyme
>3 tablespoons oil and vinegar

Combine all ingredients in a salad bowl, toss gently and serve.
>**Serves 2**

Aloe Vera Flowers x

Eat flowers raw and add to favorite sandwiches, salads. Dip flowers in favorite sauces.

Wild Greens Quiche ~

Crust

    3 cups cooked brown rice
    1 cup sesame seeds
    2 tablespoons butter
    dash of salt

Filling

    1 tablespoon butter
    1 medium onion, chopped
    8 ounces fresh mushrooms, sliced
    4 garlic cloves, crushed
    3 cups mixed wild greens (lamb's quarters, amaranth, clover, dandelion, curly dock, or plantain), washed
    1/2 teaspoon salt
    3 eggs, beaten
    1 cup milk
    1 teaspoon marjoram leaves
    1 teaspoon chopped basil leaves
    1/2 teaspoon thyme leaves

Preheat the oven to 350 degrees.

Mix the brown rice, sesame seeds, butter, and salt in a bowl and press into a 12-inch pie dish. Set aside. Melt the butter in a large skillet. Add the onion, mushrooms, garlic, greens, and salt and sauté until tender, about 10 minutes. Set aside.

Combine the eggs, milk, marjoram, basil, and thyme and sauté until tender, about 10 minutes. Combine with greens and pour into shell. Smooth top and bake for 45 minutes, or until brown and crispy on the top.

Serves 8 as a main course

Baked Purslane with Yogurt x

    1 cup fresh purslane, washed
    1 pint yogurt

Preheat the oven to 350 degrees. Mix the purslane and yogurt and pour into a greased 1-quart casserole. Bake 10 to 15 minutes or until hot.

Serves 2 as a side dish

Tabouli with Wild Greens ~

    1 cup young dandelion leaves
    1 cup daisy leaves
    1 cup red clover leaves
    1 cup chopped sheep sorrel leaves
    1/2 cup violet leaves
    1/4 cup chopped spearmint leaves
    1/4 cup chopped fresh oregano, or 2 teaspoons dried
    1 1/2 cups cooked brown rice
    1 1/2 cups cooked bulgur
    oil and vinegar to taste
    1 cup daisy petals
    1 cup violet and strawberry flowers or wild roses

Wash and drain all greens and tear into bite-size pieces. Place in a large salad bowl. Toss in the rice, bulgur, oil, vinegar. Arrange the daisies around the edge of the bowl and place the violets and strawberries or roses in the middle. Chill and serve.

    Serves 6 to 8

Lamb's Quarters Calzone x

Dough
    1/4 cup warm water
    1 tablespoon active dry yeast
    4 teaspoons honey
    1 cup soymilk
    2 eggs
    3 cups unbleached all-purpose flour
    1/4 cup lamb's quarters flour
    1 1/2 teaspoons sea salt

Filling
    1 cup pizza sauce
    1 teaspoon chopped garlic
    1 medium onion, chopped
    1 cup chopped fresh mushrooms
    1 green bell pepper, chopped
    1/2 cup chopped ripe olives
    1/4 cup chopped lamb's quarters
    12 ounces fresh mozzarella cheese, shredded
    3 tablespoons olive oil

Put the warm water into a small bowl and stir in the yeast and honey; let the mixture stand until foamy, 5 to 10 minutes. Whisk the soymilk and 1 egg into the yeast mixture.

In another large bowl, mix the flours and salt. Make a mound of flour then make a well in the middle. Gradually pour the yeast and milk mixture into the well, working the flour from the inside of the well into the liquid with your fingers. Continue until all the flour is absorbed, adding 1 tablespoon of water if needed. Knead until soft, velvety, strong, and elastic, 10 to 15 minutes. Place the dough in a lightly oiled bowl, cover with a damp cloth, and put in a warm place. Let the dough rise until doubled, about 2 hours.

Mix the ingredients for the filling and set aside.

Knead the dough briefly on a lightly floured surface. Cut the dough into 6 or 7 pieces. Shape each piece into a ball, flatten it hard with your hand, and roll it into a 1/4-inch thick oval with a rolling pin. Mix the remaining 1 egg with 1 tablespoon of water and brush the edge of each shaped oval with the mixture.

Preheat oven to 450 degrees, lightly oil a baking sheet.

Place 2 tablespoons of pizza sauce in the center of each dough oval. Add a little garlic, onion, mushrooms, green pepper, olives, and lamb's quarters to each and top with cheese. Fold over the dough to form a half-moon shape, crimping the edges with a fork. Place the calzones on the baking sheet and brush each with the oil. Bake for 20 to 25 minutes or until golden brown. Brush again with oil and serve hot.

Serves 6 to 7

Weed Pizza x

    2 tablespoons butter or olive oil
    3 cups mixed wild greens, such as clover, plantain,
        dandelion, lamb's quarters, washed
    dash of salt
    1 pound ready-made pizza dough, thawed if frozen,
        toppings of choice

Preheat the oven to 425 degrees. Heat the butter or oil in a large frying pan. Add the greens and salt, and sauté for 5 minutes or until tender. Roll out pizza dough according to package instructions. Spread the greens on the crust and sprinkle with toppings. Bake for 20 minutes or until crisp.

Makes 6 slices

Malva Neglecta and Lamb's Quarters Quiche x

    6 eggs
    1 tablespoon lemon pepper
    1 cup chocolate flavored soymilk
    1/4 cup chopped malva neglecta leaves
    1/4 cup chopped lamb's quarters
    1/4 cup grated carrot
    2 tablespoons chopped onion
    1 teaspoon chopped garlic
    1 9-inch pie shell, unbaked

Preheat the oven to 350 degrees. Process the eggs, lemon pepper, and soymilk in a blender or food processor until thoroughly blended. Stir in the malva neglecta leaves, lamb's quarters, carrot, onion, and garlic. Pour the mixture into the pie shell and bake for 35 minutes or until a knife inserted in the center comes out clean. Serve hot.
    Serves 5 to 6 as main course

Milkweed Pod Fritters *

    12 milkweed pods, washed. Note: Pick the milkweed pods when they are young and not elastic, 2 to 3 inches in length.
    1 to 1 1/2 quarts water
    1 cup cornmeal
    1 cup whole wheat flour
    1/4 cup safflower or peanut oil

Place the pods in a large bowl and cover with the water. In a separate bowl, combine the cornmeal and flour. Roll the pods in this mixture.

Heat the oil in a frying pan. Fry the pods, turning them over, until they are cooked through and brown on all sides, about 5 minutes. Drain well.
    Serves 4 as a main dish

Milkweed Casserole *

    3 cups milkweed buds, washed
    2 cups milk or water
    1 cup whole wheat bread crumbs

Preheat the oven to 350 degrees. Grease a 2-quart casserole. Place the milkweed buds and milk or water in casserole. Sprinkle the bread crumbs on top. Cook for 20 to 30 minutes, or until bubbly and brown on top.

Serves 6 as a main course

Young Milkweed and Pasta Medley ~

- 1 pound assorted pasta (rigatoni, linguine, broken into thirds, small shells, bucatino broken into thirds)
- 2 tablespoons olive oil
- 1/4 cup chopped wild onions or yellow onion
- 3 garlic cloves, coarsely chopped
- 1/4 cup dry sherry
- 2 cups milkweed shoots and leaf buds, blanched
- ground thyme or pepper, Queen Anne's lace seeds or salt
- 1/4 cup chopped fresh parsley or watercress
- 1/4 cup grated Pecorino Romano cheese (optional)

Cook the pastas until al dente; drain and set aside. Heat the oil in a frying pan and sauté the onions and garlic until lightly browned. Add the wine and cook until the liquid is reduced by half, about 9 to 10 minutes. Add the milkweed and sauté until tender, approximately 10 minutes. Add the pastas and toss, heating thoroughly. Season to taste with thyme and seeds as salt. Garnish with parsley or watercress and sprinkle with cheese.

Serves 4 as a main course

Daisy Leaves a la Mediterranean x

- 2 cups washed daisy leaves
- 2 tablespoons olive oil
- 1 clove garlic, chopped finely
- 1 teaspoon thyme

Heat frying pan and add olive oil and chopped garlic. Brown garlic and add washed daisy leaves. Stir until piping hot.

Serves 1

Mustard Greens Casserole *

    2 big bunches wintercress greens stems, leaves, and flowers
    4 to 6 tablespoons safflower oil or butter

Preheat the oven to 350 degrees. Wash and drain the wintercress. Steam until wilted, about 3 minutes. Remove the tougher stems and chop the leaves & short stems finely. Steam the chopped mixture for 15 minutes more, stirring frequently. (Keep the water used in steaming for soup stock; freeze if necessary.) Pour the greens into a casserole dish, dot with oil or butter, and bake for 8 minutes or until hot and bubbly.
    Serves 4 as a side dish

Stir-Fried Mustard Flowers *

    1 quart mustard flower heads
    1 garlic clove, chopped
    1 tablespoon olive oil

In a frying pan or wok, quickly stir-fry all ingredients over high heat until limp, about 2 to 3 minutes. Serve hot.
    Serves 4 as a side dish

Mustard Flower Buds

    4 cups mustard flower buds
    1 quart boiling water
    2 tablespoons olive oil

Wash the buds thoroughly and add to boiling water. Lower heat, and simmer 5 minutes. Drain, add oil, and serve hot.
    Serves 4 as a side dish

Vegetarian Burger x

    2 cups cooked lentils
    1/2 onion, peeled
    2 tablespoons sow thistle leaves

Steam then chop thistle leaves. Add other ingredients and mix until everything sticks together.
    Makes 6 medium-size burgers

Nettle-Falafel Burgers x

    1 cup cooked and finely chopped nettles
    1/2 cup falafel mix (see box for recipe)
    2 tablespoons olive or safflower oil

Place the nettles in a medium bowl and add the falafel mix slowly. Mix thoroughly. Add a bit of water at a time until a clay-like mixture results. Pat the mixture into 3-inch patties. Heat the oil in a medium frying pan and fry the burgers slowly until brown on both sides, about 6 minutes.

Serves 2 to 3 as a main course

Boiled Mustard Greens !

    1 quart young mustard leaves
    1 garlic clove, chopped
    1 teaspoon lemon juice
    1 teaspoon vinegar
    1 tablespoon soy butter or regular butter.

Boil the leaves for 30 minutes in a large saucepan. Drain well and add the garlic, lemon juice, vinegar, and butter. Serve hot.

Serves 2 as a side dish

Purslane Casserole !

    1 teaspoon peanut or olive oil
    4 cups purslane (including stems), washed, drained, and chopped
    1 egg, beaten, or violet leaf thickener, opposite page
    1/2 cup fine bread crumbs

Preheat the oven to 325 degrees and grease a 1 1/2-quart baking dish with oil. To the purslane add the egg or violet thickener and the bread crumbs. Turn into the baking dish, and add your favorite seasoning. Bake for 20 minutes or until piping hot.

Serves 4 as a side dish

Mexican-Style Purslane ~ (Vertalogas)

The following recipe was given to me by a Mexican friend. It is a common dish in Arizona.

1/2 pound purslane leaves and tender stems
1 teaspoon baking soda
1/2 pound pork chops or ribs
1/4 medium onion, chopped
1 garlic clove, chopped
1 medium tomato, chopped
1/2 teaspoon chicken broth

Place the purslane in boiling water with a pinch of baking soda and cook until tender, about 2 to 3 minutes.

In a large skillet, brown the meat on both sides and set aside. In the same pan sauté' the onion, garlic, and tomato until tender. Stir in the broth, purslane, and meat, and simmer for 15 to 20 minutes or until meat is done.
Serves 5 as a main course

Violet Leaf Thickener x

a handful of violet leaves
1/4 cup boiling water f

Press leaves down under boiling water. Do this slowly and continue for about 20 minutes. Drain. Green water is slightly gooey-or thickened. Use as water substitute. For cassaroles, layer whole violet leaves throughout casserole. Water from other ingredients will draw thickener from the edible violet leaves.

Wild Lettuce Stir-Fry x

1 cup wild lettuce leaves
1/2 cup wild lettuce buds and flowers
1/2 cup red cabbage
1 clove garlic, pressed
1 teaspoon wild thyme
3 tablespoons oil

Stir-fry all ingredients the oil, serve hot.
Serves 4

Brown Rice and Lamb's Quarter Sprouts x

    1 cup lamb's quarter sprouts (sprout seeds only 1 1/2 to 2 days)
    2 tablespoons corn oil or peanut oil
    1 tablespoon chopped scallions or chives or onions
    1/4 cup thin sliced canned or fresh mushrooms
    1 cup cooked brown rice

Heat oil in heavy fry pan or wok. Add onion, mushrooms. Stir fry until lightly brown. Add soy sauce, rice and sprouts. Stir fry for 3 to 4 minutes. Cover and simmer for 5 more minutes.
    Serves 3 to 4

Wild Malva Rice Roll-Ups x

    1 cup water
    12 large malva neglecta leaves
    1 cup cooked wild rice
    1/2 teaspoon thyme
    1/2 teaspoon vegetable salt (of your choice)
    small casserole dish, greased
    1/4 cup salsa to your taste

    Bring water to a boil. Wilt malva leaves by gingerly dipping each leaf separately into boiling water. Rearrange leaf on a paper towel to drain. Place wild rice in a bowl. Stir in thyme and 1/4 teaspoon vegetable salt.
    Take a teaspoon of mixture and place in the middle of the leaf. Roll the leaf and mixture up to stem, then tuck stem end in to secure. Place whole roll-up face down in casserole dish. Sprinkle rest of vegetable salt over roll-up and bake in oven for 20 minutes at 350 degrees. Pour salsa over each and serve hot.
    Serves 2

Malva Mexican Style

    1 cup large Malva leaves
    1/2 cup salsa sauce or spaghetti sauce
    1 cup refried beans1 ice cube tray

Place ice cube tray on counter. Drip salsa sauce in tray as evenly as you can. Place a leaf into cube section over salsa. Take a teaspoon and place a spoonful of refried beans into the leaf. Place the ice cube tray into the freezer and freeze solid.

Pop sections out into a freezer bag with a ziplock top. When you cook a casserole, take a few ready-made sections out and place in a bean pot. Cook 350° for 20 minutes. Delicious!

Makes 1 tray full of stuffed Malva leaves

Stuffed Malva leaves. Cook freshly rolled or freeze in ice cube trays for future meals. Store in plastic bags.

Willie Whitefeather and Linda.
Note his handmade shirt, moccasins and flute.

Muffins, Breads, Biscuits & Pancakes

Basic Mix x

    4 1/4 cups whole wheat flour
    4 1/4 cups amaranth or triticale flour (see Note)
    5 tablespoons baking powder
    1 teaspoon baking soda
    1 tablespoon ground dried Queen Anne's lace seeds
    2 teaspoons cream of tartar
    1 1/2 cups instant dry milk
    1 1/2 cups olive or peanut oil

In a large bowl, sift together the flours, baking powder, baking soda, seeds, cream of tartar, and dry milk. Blend well. With a wooden spoon, stir in the oil until evenly distributed; the mixture will be only slightly damp. Put the mix in a large, airtight container and label. Store in a cool, dry place. Use within 10 to 12 weeks.

Makes 11 cups, enough for several batches of crackers.

Note: Triticale flour is made from a high-protein grain that is a hybrid between wheat and rye. The flour is available in some natural foods stores.

Amaranth Vegetable Bread x

    1 package active dry yeast
    1/4 cup warm water
    6 cups all-purpose flour or 10 cups whole wheat flour
    4 cups cooked, drained amaranth greens

Dissolve the yeast in the warm water. Place flour in a large bowl and mix in the yeast. Add the greens and mix by hand, kneading the dough for 3 minutes. Cover and let rise 1 hour, or until doubled in bulk.

Preheat the oven to 400 degrees. Punch dough down and form into 2 loaves. Place on a baking sheet and let rise again until doubled, about 45 minutes. Bake loaves for 40 minutes, or until they sound hollow when tapped. Serve hot. These loaves freeze well.

Makes 2 loaves

Mean Green Bread x

    2 tablespoons active dry yeast
    3 tablespoons honey
    2 1/2 to 2 3/4 cups warm water
    4 cups whole wheat flour
    2 cups fresh greens (dandelion, plantain, clover, lamb's quarters)
    1 teaspoon salt
    2 tablespoons cold-pressed safflower oil
    4 to 6 cups flour mixture half lamb's quarters and half rice flour, or clover and whole wheat flour

Mix the yeast, honey, and 2 cups warm water in a large bowl and let it sit for a few minutes until foamy. Add the whole wheat flour and beat for 2 to 3 minutes. Cover the mixture and let set until doubled in size. (approximately 1 hour).

Meanwhile, in a bowl combine the greens, remaining warm water, salt, and oil. Add to the flour a little at a time and blend, kneading for about 5 minutes. Grease a large bowl and let the dough rise until doubled in size, about 50 minutes.

Punch the dough down and let rise another 40 to 50 minutes.

Preheat the oven to 400 degrees. Bake for 35 to 40 minutes or until you hear a hollow sound when you tap the top with the flat side of a butter knife. Cool for 30 minutes before serving.

    Makes 1 large loaf

Wild Malva Rolls x

    1 cup water
    12 large malva neglecta leaves, washed
    1 cup cooked wild rice
    1/2 teaspoon dried thyme leaves
    1/2 teaspoon salt

Preheat the oven to 350 degrees and grease a 1-quart casserole dish.

In a small saucepan, bring the water to a boil. Wilt the malva neglecta leaves by carefully dipping each leaf separately into the boiling water. Place the leaves on a paper towel to drain.

Place the rice in a bowl. Stir in the thyme and half the salt. Place 1 teaspoon of the mixture in the middle of a leaf. Fold the sides toward center and roll from stem to tip of leaf.

Place rolls seam side down in the casserole dish. Repeat, making the remaining rolls.

Sprinkle remaining salt over rolls and bake 20 minutes or until bubbly hot. Serve piping hot.

Serves 2 as a main course

Italian Clover Rolls *

    1 cup clover flour
    3 cups whole wheat flour
    1/2 cup clover leaves
    1/8 cup vegetable oil
    2 garlic cloves, chopped
    1 6-ounce can tomato paste
    garlic powder

Preheat the oven to 350 degrees. Combine the flours in a large bowl. Add the clover leaves, oil, garlic, and tomato paste. Using your hands, knead thoroughly, adding a small amount of water until a clay-like consistency is obtained.

Roll golf-ball size pieces of dough in your palms and place on an oiled baking sheet. Bake for approximately 20 minutes or until golden brown. Serve hot, sprinkled with garlic powder.

Makes 12 to 14 rolls

Crabgrass Muffins !

    1 cup enriched flour
    1 cup crabgrass flour
    2 teaspoon baking soda
    2 teaspoons ginger (optional as a variation)
    3/4 cup water
    2 eggs (egg replacement used by author)
    1 teaspoon pure vanilla
    1/4 cup sunflower or canola oil
    1/2 cup raisins
    2 teaspoon baking soda

Preheat oven to 350 degrees.

Place flours and baking soda in bowl, mix in water, eggs, vanilla, & oil. Fold in raisins thoroughly. Fill muffin tins 1/2 full or pour in 8-inch square baking pan.

Bake 20 to 25 minutes. Let cool and remove from pan.

Makes 6 muffins

Crabgrass Crackers x

    1 cup basic mix (Page 225)
    1 cup crabgrass flour
    1 teaspoon salt
    1/2 cup water

Preheat the oven to 400 degrees. Lightly grease a baking sheet. In a medium bowl, combine the mix, flour, and salt. Add water to form a dough. Knead about 12 times, or until the dough is smooth. Shape into pencil-like strands 1/2 inch thick, and place on baking sheet. Roll out as a single sheet onto the baking sheet. Use a fork to perforate the sections. Bake for about 5 minutes or until crisp.

Variations: Use another flour, such as from malva, mixed grain, or lamb's quarters. Substitute goat's milk or cow's milk for water.

Makes 24 to 36 crackers

Wild Waffles ~

<u>Dry Mix</u> 1 cup rice flour
    1 cup barley flour
    1 cup buckwheat flour
    1 cup old-fashioned rolled oats
    1/4 cup finely chopped lamb's quarters
    1 cup mixed dried greens (clover, dandelion, violet
        leaves, curly dock, & amaranth)
    4 tablespoons baking powder
    1 1/2 teaspoons salt

Combine the flours, oats, lamb's quarters, mixed greens, baking powder, and salt. Store in the refrigerator.

Makes 4 cups

<u>Batter</u> 2 cups dry mix
    2 cups soymilk, cow's milk, or water
    4 tablespoons butter or cold-pressed oil

Combine the dry mix, liquid, and butter or oil. Grease a waffle iron well. Spoon about 3/4 cup batter onto hot iron. Cook each waffle until it is brown on both sides, about 6 minutes. Serve with maple syrup or honey butter.

Serves 6 to 8

Navaho Lamb's Quarters Fry-Bread ~

1 cup whole wheat flour
1 cup lamb's quarters flour
2 teaspoons baking powder (optional)
1/2 cup reconstituted nonfat dry milk or water
2 teaspoons safflower oil
1/2 cup ice water
Oil for frying

Combine the flours and baking powder (if using). Add the milk or water and oil, and using your hand, mix into a granular mixture. Add the ice water and form into a ball. Put a towel over the bowl and let the dough rise in a warm place for 2 hours. (Even without baking powder, the dough rises about 1/2 inch.)

Cut the dough into 2 parts, roll out each part until 1/4 inch thick, then score parallel lines 1/2 inch apart almost through each piece. Heat oil in a deep pan until very hot, then add 1 piece of dough. Deep-fry until both sides are brown, approximately 2 minutes. Drain on paper towels and repeat for the other piece of dough. Serve hot. Can be frozen and reheated in a microwave.

Makes 2 breads

Mallow Vegetable Bread ~

5 cups warm water
12 to 14 cups whole wheat flour, ground fresh from whole wheat berries, if possible
2/3 cup olive oil
2/3 cup rice syrup
2 teaspoons sea salt
3 tablespoons active dry yeast
3 tablespoons vitamin C crystals (for dough consistency), vailable in health food stores.
2 medium onions, chopped
1/2 cup chopped mallow
3 medium carrots, chopped
4 garlic cloves, chopped

In a large bowl, combine the flour, water, oil, rice syrup, salt, yeast, and vitamin C crystals and mix thoroughly. Add the onions, mallow, carrots, and garlic. Mix thoroughly by hand, then knead for 10 minutes. Form into loaves, place in greased pans and cover. Let rise until dough rises above the pans, about 1 hour.

Preheat oven to 350 degrees. Bake loaves for 35 to 40 minutes, or until they sound hollow when tapped.

Makes 2 loaves

Phragmities Pancakes !

1 cup phragmities flour
water (approximately 1/4 cup)
oil for frying

Put the flour in a medium bowl, slowly add enough water to form a paste. Mold the paste into a pancake form. Grease a skillet and fry the cake slowly until golden brown and crispy, about 6 minutes.

Serves 1

Variation: Preheat oven to 350 degrees. Grease a baking sheet and bake the cake for 15 minutes or until brown and puffy.

Phragmities Drop Biscuits *

1 cup phragmities flour
1 cup wheat bran or whole wheat flour
2 tablespoons safflower oil
3 tablespoons baking powder
approximately 1/2 to 3/4 cups water

Preheat oven to 350 degrees. Blend the flours and oil. Add just enough water to make a paste. Spoon silver-dollar-size portions onto a greased baking sheet and bake for 10 to 15 minutes or until crisp.

Yields 18 biscuits

Variation: Add 1/2 cup honey to the recipe

Phragmities Pita Bread x

1 cup phragmities flour
1 cup whole wheat flour
1 teaspoon baking powder
1/4 teaspoon olive oil
approximately 1/2 to 3/4 cups water

Preheat the oven to 350 degrees. Combine the flours, baking powder, and oil in a large bowl. Mix slowly, adding water a little at a time until mixture is of bread dough consistency. Knead until smooth, then let the mixture sit, covered, for 10 minutes.

Form 1 inch balls into individual patties ¼ inch thick and put on a greased baking sheet. Bake until the edges of the pita breads are brown and crispy, about 30 minutes.

Makes 12 pita breads

Green Pancakes (made from any vegetable flour) !
1/4 cup flour
small amount of water

Add a tiny amount of water very slowly to the flour to form a paste. Make two 2" in diameter pancakes. These are powerful pancakes, so you want to make them small. Fry slowly on a greased fry pan or cooking sheet, or bake in the oven as you would a cookie. Instant energy boost!

Makes two 2" pancakes

Nutsedge Crackers ~
1 cup Basic Mix (page 225)
1 cup nutsedge flour or oat flour
1/2 cup water

Preheat oven to 400 degrees. Lightly grease a baking sheet. In a medium-sized bowl, combine Basic Mix and flour. Add water to form dough. Knead about 12 times, until dough is smooth. Shape into pencil-like strands 1/2" thick or roll out flat onto baking sheet. Bake about 5 minutes until crisp. Cut into small pieces while warm.

Makes 24 to 36.

Variations: Use any flour you have such as malva, mixed grain or lamb's quarters in place of nutsedge or oat. Use goat's milk or regular milk in place of water.

Breads and Flour

The following will make excellent flour. Refer to pages 35-40 for information on drying, grinding and storage.

- plantain leaves
- clover leaves
- dandelion leaves
- strawberry leaves
- cattail fluff, root
- curley dock seeds and root
- nutsedge (grainy top of plant)
- Lamb's quarter leaves
- pigweed leaves
- bulrush pollen
- thistle root
- inner bark of birch bark

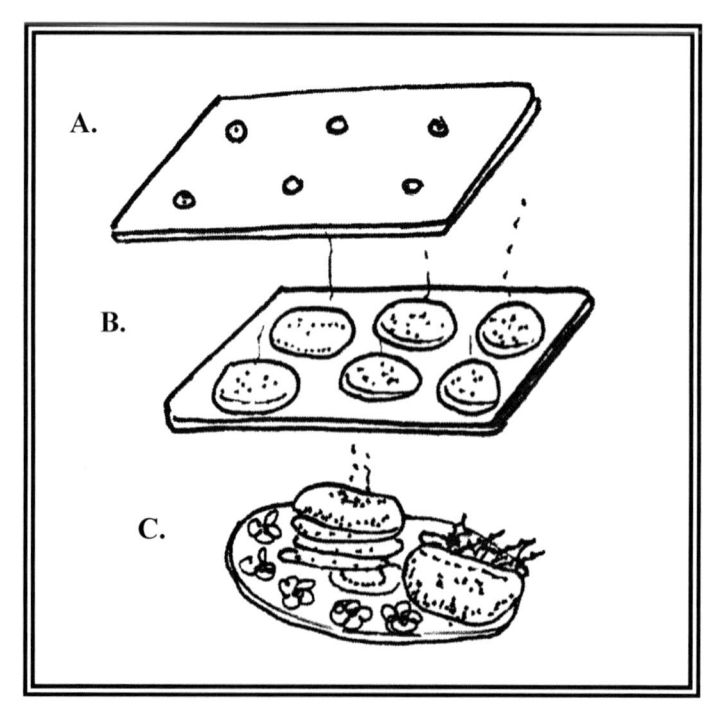

Homemade Wild Pita Bread.
A. Drop round balls of dough
B. Bake until edges are brown
C. Slice and fill pocket

Sweets

Lamb's Quarters Granola !

1 cup dried lamb's quarters seeds
3 cups old fashioned rolled oats
1 egg
1/2 teaspoon ground cinnamon
dash of ground cloves
1/4 teaspoon salt
1 1/2 cups water
3 tablespoons maple syrup, honey, or rice syrup

Combine the seeds, oats, eggs, and spices in a large bowl. In a large saucepan, bring the water and sweetener to a boil. Add the oat mixture a little at a time while stirring. Reduce the heat to low and use a fork to mix the granola until it is moist, but not wet. Add additional water, if needed, to make moist. Press layer into greased pans or pie plates. Allow to cool.
Serves 4

Variation: Add 3/4 cup raisins or 1 1/2 cups diced apples to the mixture.

Baked Prickly Pears ~

6 large prickly pear fruits
1/2 cup honey
1/2 teaspoon ground cinnamon
1/4 teaspoon grated lemon peel
4 tablespoons butter or 2 teaspoons safflower oil

Preheat the oven to 350 degrees.
Use a fork to hold the fruits while peeling to prevent the hair-like barbs from going into your fingers. After the fruits are peeled, cut in half, and remove the seeds. Grease a 6" baking dish. Line the bottom of the dish with the fruit. In a separate bowl, combine the honey, cinnamon, lemon peel, and butter or oil and drizzle over the fruit. Bake until tender, 20 to 30 minutes. Serve warm.
Serves 3

Meadowsweet Apples !

    4 large baking apples, washed and cored
    1 cup meadowsweet flowers, washed
    1/2 cup maple syrup

Preheat the oven to 350 degrees. Grease a 1-quart glass casserole dish. Place the apples in the dish and put flowers in core of each apple. Drizzle the maple syrup over the apples and bake for 30 minutes, or until soft.
    Serves 4

Malva Neglecta Oatmeal Cookies x

    1 egg
    1/2 cup honey
    1/4 cup vegetable oil
    3/4 cup finely chopped malva neglecta
    1 cup old-fashioned rolled oats
    2 teaspoons baking powder
    1 cup whole wheat flour
    1/2 cup sunflower seeds
    1 1/2 teaspoons vanilla extract
    2 teaspoons water
    1/2 cup raisins

Preheat the oven to 350 degrees. Mix the egg, honey, and oil in a blender. Stir in the malva neglecta.
    In a large bowl, blend the oats, baking powder, flour, sunflower seeds, vanilla, water, and raisins. Add the malva mixture and stir well. Drop by teaspoons onto a greased baking sheet. Bake for 12 to 15 minutes, or until golden brown on both sides. Cool on a rack.
    Makes 3 dozen cookies

Plantain Raisin Cookies x

    2 cups whole wheat flour
    3/4 cup dried or fresh plantain seeds
    4 tablespoons baking powder
    2 tablespoons molasses
    1/2 cup carob-covered raisins

Preheat the oven to 350 degrees. Mix all the ingredients in a big bowl. Add water slowly to form a thick, paste-like batter. Roll a pinch of dough between your palms and press onto a greased baking sheet. Continue to form cookies, then bake for 15 minutes or until golden brown.

Makes 2 dozen small cookies

Red Sumac Gelatin x

3 cups sumac berries
2 cups hot water
1 envelope unflavored gelatin
sumac berries for garnish

Make a concentrated sumac tea by putting the berries into a medium saucepan. Add the hot water and steep for 15 to 20 minutes. Strain. Follow the package directions for the gelatin, using the sumac tea in place of water. Garnish with a few berries.

Serves 4

Pine Popsicles *

gather the pine twigs, then strip the needles off the twigs. Use the twigs for popsicle sticks!
1 1/2 cups pine needles
2 cups boiling water

Put the needles in a small pot and pour in boiling water. Cover and steep for 10 to 15 minutes. Stir the needles once or twice to liberate more pine taste. Cool, then strain and pour into popsicle molds. Add pine twigs for sticks and freeze. (Some people put a dash of honey into the mold before pouring in the liquid for a sweet flavor.)

Makes 6 popsicles

Aloe Vera Ice Cubes x

fresh aloe vera leaves

Place aloe vera leaves on the flat side and skim the skin off with a sharp knife. Scrape the gel with a spoon, holding on to end of leaf. Pour into ice cube trays and freeze.

Mint Popsicles x

>1 handful mint leaves, washed
>mint leaves for garnish

Place the mint in a pot and add boiling water to cover. Cover the pot tightly and steep for 10-15 minutes. Strain and pour the liquid into popsicle molds. Place a leaf in each mold and freeze.
>**Makes 6 popsicles**

Children love this refreshing treat. If you don't have popsicle molds, freeze in ice cube trays, adding a mint leaf to each cube. This is delicious in iced tea and other beverages.

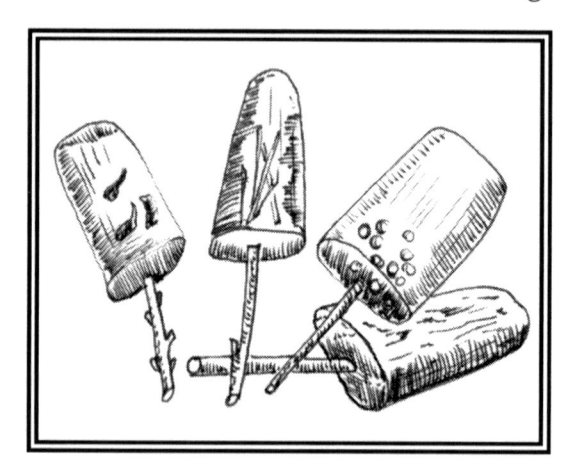

Rose Petal Candy x

>4 tablespoons butter
>2 cups rose petals
>1/2 cup confectioners sugar

Melt the butter in a large frying pan. Fold the rose petals, one at a time, and coat with butter, flipping them over for a minute or so. The petals sometimes puff up. Remove as soon as crisp and brown, then drain on a paper towel. Roll fried petals in sugar and refrigerate until ready to serve.
>**Makes enough for 4 people**

Fresh or dried, rose petals make delicious candy. Reconstitute dried petals by soaking them in water for an hour or so. Drain and use as fresh.

Wintergreen Candy x

    3 cups dried wintergreen leaves
    2 cups water
    2 cups brown sugar
    1 tablespoon distilled white vinegar

Slowly boil the leaves in the water for 20 minutes to make a strong tea. Strain. Add 6 tablespoons of tea to the brown sugar and boil until hard ball stage. Add the vinegar and boil for an additional minute or so. Pour the mixture onto a well-greased platter. It will harden as it cools and can be broken easily into pieces of candy.

    Makes enough for 4

Blueberry Crumb Pudding x

Topping

    1/4 cup sugar
    1/3 cup flour
    1/2 teaspoon cinnamon
    1 tablespoon margarine
    1 tablespoon shortening

Mix together and save for crumb topping.

Pudding

    1/4 cup shortening
    1/4 cup sugar
    1 1/2 teaspoon baking powder
    1 egg
    1 cup sifted flour
    1/8 teaspoon salt
    1/3 cup milk
    1/4 teaspoon vanilla
    2 cups blueberries

Cream shortening and sugar. Add the egg, flour, baking powder, salt and milk. Spread this in an 8" square pan and spread 2 cups blueberries over this. Top with crumb spread. Bake at 350 degrees for 30 minutes.

    Serves 4

Wild Plantain Cookies ~

    2 cups whole wheat flour
    3/4 cup plantain seeds (dried or fresh)
    4 tablespoons baking powder
    2 tablespoons molasses
    1/2 cup carob covered raisins

Mix all ingredients well in a big bowl. Add tepid water slowly to form a thick, claylike paste. To form cookies, roll a pinch of dough between your palms and press onto a greased cookie sheet. Bake 15 minutes at 350 degrees or until golden brown.
    Makes 18 to 20

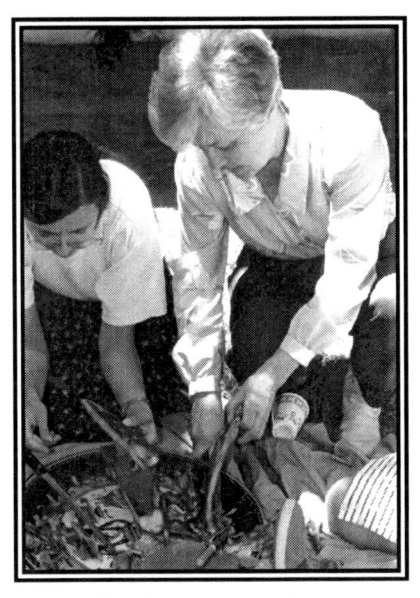

Right: Apprentice Althea Dixon slices and dices a nopales cactus.

Apprentice Hazel McManus presents a completely edible centerpiece and meal.

Apprentice Tracy Alderman Dockery teaching homeschoolers, Arizona.

Apprentice Shirlee Pettipiece.

Teas & Other Hot & Cold Drinks

Teas are an excellent way to develop your taste for wild foods. The wild ingredients in these teas are fresh unless otherwise stated. You may wish to use a tea infuser rather than strain the tea. As your taste for wild foods develops, you may even wish to eat the brewing grounds after drinking the tea.

Rule of Thumb: When you have fresh tea leaves, measure 2 cups of water per handful of leaves. Bring to a boil, cover and turn off heat. Let steep for 5-10 minutes. For dry ingredients (which cook more quickly), measure 1 cup of water per handful. Bring to a boil, cover and turn off heat. Let steep for 5 minutes. Generally, the flavor is more intense with dry herbs.

Aloe Vera Morning Tonic x

1 aloe vera ice cube (see Note)
3/4 cup fresh orange juice

Combine ice cube and orange juice in a glass. Let cube melt, then drink.
Serves 1

Note: To make cubes, place an aloe vera leaf on its flat side and peel the rounded skin off with a sharp knife. Scrape the gel with a spoon, holding on to one end of the leaf. Pour the gel into ice cube trays and freeze.

Balsam Tea !

1 inch tips balsam
1 cup water
honey (optional)

Place balsam tips in a small saucepan and add the water. Cover. Bring to a boil, then lower the heat and simmer for 10 minutes. Strain tea if desired, add water to taste if tea is too strong or sweeten with honey if desired.
Serves 1

Blackberry Bramble Tea !

 1 6 inch blackberry shoot
 1 cup boiling water
 honey (optional)

Peel the shoot until most of the bark is off. Break into several sections to expose more of the surface. Place in a teapot, add water, cover pot and steep for 5 minutes. Strain if desired. Sweeten with honey, if desired, and serve hot or cold.
 Serves 1

Balsam Toddy x

 1 to 3 inches balsam twigs with needles
 2 cups soymilk
 dash of ground cinnamon

Place balsam twigs in a teapot. Bring the soymilk almost to a boil, remove from heat, and pour over balsam. Steep, covered, for 5 minutes. Strain, then add cinnamon and serve piping hot.
 Serves 1 or 2

Goldenrod Tea !

 1 sprig goldenrod flowers
 2 cups boiling water
 honey (optional)

Place flowers in a teapot. Add water, cover pot, and steep for 5 minutes. Sweeten with honey if desired. Serve hot.
 Serves 1 or 2

Chamomile Tea !

 1/4 cup chamomile leaves and flowers, washed
 2 cups water
 honey (optional)

Place chamomile in a small saucepan. Add water, bring to a boil, turn off heat, and cover. Steep 10 minutes, then strain and serve with honey if desired.
 Serves 1 to 2

Wild Greens Tonic x

This tonic will keep in the refrigerator for 2 days or so. I freeze the brew as ice cubes for use in recipes or for a cup of tea.

- 1 cup willow leaves, washed
- 1 cup mint leaves, washed
- 1/2 cup red clover buds
- 2 quarts water

In a blender of food processor, blend all ingredients until green and smooth. Serve hot or cold.

Serves 8

Dandelion Root Coffee or Chicory Coffee !

- 1 teaspoon dandelion root or chicory root powder (see page 76)
- 1 cup water

In a medium saucepan, combine the powder and water. Bring to a boil and simmer, covered, for 3 to 4 minutes. Drink as a coffee substitute.

Serves 1

Hot Meadowsweet Tea !

- 4 white meadowsweet flower spires
- 4 to 6 meadowsweet leaves
- 1 cup water

Bruise the leaves and flowers, then pour boiling water over them and cover for 7 minutes. Strain. Serve hot, or add 1 teaspoon lemon juice and serve cold.

Serves 1

Maple Tea !

- 1/2 cup maple seeds, clipped from the wing
- 1 cup boiling water
- 2 teaspoons maple syrup (optional)

Place seeds in a teapot, add water, and steep for 5 minutes. Sweeten with maple syrup if desired.

Serves 1

Mullein Root Tea !

Mullein root is an antihistamine. When you feel a cold coming on, simmer the root in water and drink the liquid. The root can be used several times to make tea. After five or six uses, bruise the root with a hammer, or cut slits in the sides to release more nutrients.

>1 large mullein root
>honey (optional)

Scrub the root well and cut slits in several places with a strong knife. Place the root in a non-corrosive pot. Add water to cover, bring to a boil, and simmer, covered, for 5 minutes. Sweeten with honey if desired. Hang the root up to dry.
Serves 1

Willow "Aspirin" Tea !

For pain relief, chew the willow twigs raw as well.

>1 6-inch willow tendril, including leaves and catkins
>1 cup boiling water
>honey (optional)

Using a knife edge, score and remove the bark. Slice the bark, leaves, and catkins into a teapot. Break up the twig into several sections and add to the teapot. Add water, cover, and steep for 5 minutes. Strain, and sweeten with honey if desired.
Serves 1

Pine Needle Tea !

A pine tree is filled with vitamin C, so pine tea is delicious and nutritious. Cold or hot, the flavor is wild. Perhaps in time you will develop a taste for *raw* pine needles. They are edible and so is the bark. Your backyard tree might hold several hundred cups of tea—perhaps thousands. Seasons make no difference; pine is green and succulent all year long.

>1 cup fresh pine needles
>1 cup boiling water
>honey (optional)

Put the needles in a teapot and pour in the water. Steep for 5 to 7 minutes, then strain. Sweeten with honey if desired.
Serves 1

Wintergreen Tea !

Wintergreen tea is scrumptious. With this tea you can comfortably pass a cold winter's night. Gather the flat, shiny leaves and spread them on a screen to dry in a warm, airy spot. The back window ledge of a car is good. Let them dry slowly in the sun and watch that they don't dry too fast. This technique does not work in August when the sun is too hot. I use a hand grinder or old meat grinder to obtain the powder for this tea.

> 2 teaspoons ground dried wintergreen leaves
> 1 cup boiling water
> honey (optional)

Put wintergreen into a cup and add water. Steep for 5 minutes. Sweeten with honey if desired.
Serves 1

Violet Tea !

> a handful of violet flowers
> 2 cups of water

Pick flowers after dew has evaporated and dry on a screen for a day. Add flowers to the water and cover. Steep for 10 minutes. Skim flowers out or eat them!
Serves 2

A Choice of Blend of Herbs for Tea !

Gather a combination of lawn herbs. Examples: thyme, mint, red sorrel, yarrow, spearmint, goldenrod, rose hips, among a few. Arrange the herbs so the stems are the same length and bundle them together. Hang the bundle high in a warm kitchen. When the leaves are dry enough to crumble, rub the bundle between your palms over a tray. Then rub that mixture through a sieve. You can bottle this Fine Blend and store for future exquisite tea! Also, one tablespoon is adequate for a cup of soup.

Blackberry Bramble Wine !

> 1 gallon blackberry shoots or tips
> 1 pound raisins, chopped
> 1 gallon boiling water
> 3 1/2 cups sugar or 2 cups honey
> 1 tablespoon all-purpose wine yeast

Wash the blackberry shoots. Using scissors, clip shoots into 1-inch pieces and place in a sterilized 5-gallon container. Add the raisins, then cover with boiling water. Cover tightly. The next day, strain liquid and return it to the container. Add the sugar or honey and stir well. Prepare the yeast according to the package directions. When bubbly, add the yeast mixture. Cover tightly. In a few hours the mixture will begin to foam. Stir occasionally, until fermentation slows. When bubbling ceases (in about 6 weeks), siphon the liquid into sterilized bottles. Cork and let sit in a dark place. Serve in a year's time.

Makes about 5 bottles

Daisy Petal Wine !

    1 gallon daisy heads with petals (no stems), washed
    1 gallon boiling water
    3 pounds sugar or 1 1/2 pounds honey
    1 package all-purpose wine yeast

Pack the daisy heads into a sterilized 2-gallon crock. Pour the boiling water over the heads and stir. Cover tightly. Let the mixture stand for 1 1/2 weeks. Strain.

Bring the liquid to a boil and remove from the heat. Transfer to a sterilized 4-gallon container. Stir in the sugar or honey until it is dissolved. Let cool. Meanwhile, prepare yeast according to package instructions. Add to liquid and cover tightly. Let the mixture sit for a week while it foams. Let it sit until the fermentation clears, about another 6 weeks. Siphon into sterilized bottles and store for 4 to 6 weeks before serving.

Makes about 5 bottles.

Dandelion Wine !

    1 gallon dandelion flowers
    4 oranges
    1 gallon water
    5 pounds sugar
    1 package all-purpose wine yeast

Remove the green parts and soak the flowers in boiling water. Peel oranges and save the juice. Pour boiling water over flowers and let stand for 2 days. Add orange peels, bring to a boil, then let simmer for 15 minutes. Strain liquid and parts through a fine strainer or use a jelly-bag. Add sugar to strained liquid and cool to room temperature. Add the juice of the oranges and yeast. Pour into a fermenting jug fitted with an airlock. When fermentation clears, siphon into sterilized bottles. Store at least 6 months before serving.

Makes 5 quarts

Prickly Pear Wine x

    1 dozen prickly pear buds
    2 quarts boiling water
    4 cups honey
    1 package all-purpose wine yeast

Hold the prickly pear buds with forceps in one hand and peel under water. Place pears in a sterilized 1-gallon crock and pour in boiling water. Cover tightly. Place mixture in a warm place for 4 or 5 days or until mixture "turns."

Strain liquid and return to gallon crock. Add honey. Prepare the yeast according to package directions and stir into the mixture. Cover it tightly, and let it sit in a dark warm spot for 1 more week or until foaming clears. If desired, transfer to a carboy fitted with an airlock. Let it sit until the fermentation ceases in about 6 weeks.

Siphon into sterilized bottles and cork tightly. Let rest for about 8 months.

Makes about 3 bottles.

Rose Petal Wine !

    1 quart rose petals, washed
    2 quarts water
    2 cups honey
    1 package all-purpose wine yeast

Simmer the rose petals in water for 5 minutes, stirring occasionally. Pour the liquid and petals into a sterilized 2-gallon glass container. Cover tightly and let the mixture sit for 1 week. The petals and water will turn almost rancid.

Strain the mixture and put the liquid in a clean pot. Bring to a boil, then turn off heat. Transfer to a sterilized 2-gallon container. While mixture is very warm, add the honey. Let it cool. Meanwhile, prepare the yeast according to package directions. When the mixture has cooled, add the yeast mixture, stir once, and cover tightly. If desired, after 1 week, transfer the mixture to a fermenting jug fitted with an airlock. When the fermentation ceases, in about 6 weeks, siphon the wine into sterilized bottles and cork tightly. Let rest for 4 months.

Makes about 3 bottles

Sugar Maple Twig Wine !

    1 gallon tightly packed sugar maple twigs, washed
    1 gallon boiling water
    4 cups honey
    1 package all-purpose wine yeast

Pack the twigs into a casserole. Pour in the water and simmer for 20 minutes. Transfer to a sterilized 2-gallon crock. Cover tightly. Let sit for 3 days.

Strain the liquid and add honey. Stir well. Prepare the yeast according to package directions and add to the mixture. Stir, then cover tightly. Mixture will foam for 1 week. If desired, at this point transfer to a sterilized fermenting jug fitted with an airlock. Keep covered until fermentation stops, in about 6 weeks. Siphon into sterilized bottles and cork. Store in cellar for 3 to 4 months before serving.

Makes about 5 bottles

Preserving Wild Foods

Dandelion Jelly x

    Petals from 40 dandelion flowers
    3 cups boiling water
    1 package commercial pectin
    3 3/4 cups sugar or 1 3/4 cups honey

    Wash the petals and add them to the water. Steep for 10 minutes, then strain. Return the juice to the heat, bring to a boil, and immediately add the pectin. Boil mixture for 1 minute, then add sugar or honey and simmer until it makes a hard ball (260 degrees). Pour into hot, clean pint jars and seal with wax or process in hot water bath for 20 minutes.
    Makes 2 pints

Cactus Lemon Marmalade x

    1/4 cup thin lemon slices
    1 cup prickly pear cactus pulp (Split the fruit and
        remove the tips & seedy centers.)
    1/4 cup honey

    Put the lemons in a small saucepan and cover with water. Soak overnight. The next morning add the cactus and honey and cook until the mixture is thick. Pour into a hot pint jar and cover with sealing wax or process in a hot water bath for 20 minutes.
    Makes 1 pint

Saguaro Cactus Jelly x

    16 to 24 saguaro fruits
    1 package powdered pectin
    1/4 cup lemon juice
    2 1/2 cups honey

    Split the fruit and remove the tips & seedy centers. Place the pulp in a pot, cover with water, and simmer until well-cooked, about 15 minutes. Strain the pulp using a fine colander. In a large pot, combine 3 1/4 cups saguaro puree, pectin, lemon juice,

and honey. Boil for 1 minute, stirring constantly, then pour into scalded pint jars. Cover with sealing wax or process in a hot water bath for 20 minutes.
Makes 4 pints

Rose Petal Syrup !

    4 cups rose petals
    1 quart water

Put the petals in a large saucepan and cover with water. Simmer for 20 to 25 minutes, stirring often, until it is thickened to syrup consistency. Cool, bottle, and refrigerate.
Makes 1 quart

Rose Hip Jam !

    4 cups rose hips, washed
    4 cups water
    2 cups honey

Cook the hips in the water until tender, about 15 minutes. Put the hips through a meat grinder, food processor, or coarse sieve, then push the pulp through a fine sieve to obtain a smooth puree. Add 1/2 cup honey for each cup of pulp and cook until thick. Pour into scalded pint jars and seal with sealing wax or process in a hot water bath for 20 minutes.
Makes 4 pints

Queen Anne's Lace Salt !

This salt substitute lasts for years and years. Caution: Be 100 percent certain on your identification. The fuzzy hairs on the stem of Queen Anne's lace are dry at this stage. Running your finger down the dried stem feels like sandpaper.

Collect Queen Anne's lace seeds in the early fall, when flowers have "birdcaged" and turn a light brown. When the seeds mature, the entire head naturally dries up. Pull on the dried seed top; the seeds are easily extracted in your hand, even without taking the "birdcage" off the stem. The remaining seeds will propagate new growth next season.

Place the seeds on a tray and put into a 250 degree oven for 5 minutes. You may put them on a tray on top of the stove's pilot light until completely dry. Bottle in glass jars for long-term storage.

Caution: Pregnant women should not use these seeds: medicinal field guides show that the seeds have abortive tendencies.

Herbal Vinegars !

Fantastic herbal vinegars can be made from the blossoms of red or white clover, roses, catnip, mint, goldenrod, blackberries, raspberries, violets, and lamb's quarters. You might enjoy adding other flavors, such as thyme, garlic, or Queen Anne's lace. My favorite vinegar is a combination of greens and spices. Experiment with different flavors to find your favorite. These vinegars make wonderful gifts.

Wash and drain the greens and flowers and place in a 2-quart glass container. Bring vinegar to a boil, then pour it slowly over the greens and flowers. Cap and store the container in a high place, out of direct sunlight. The vinegar will cook the greens a bit in the warmth. Shake it every day for best results. After 2 weeks, strain the mixture, cap tightly, date and label the container.

Makes 2 quarts

Lemon Pepper x

    1 peel from lemon, scrubbed
    1 2-ounce container garlic salt
    1 3-ounce container celery salt
    1 2-ounce container coarse black pepper

Dry lemon peel thoroughly, about 3 days on a plate or 1 to 1 1/2 hours in an 250-degree oven. Don't let it get brown. Pulverize dried peel in blender until powdered. Add the rest of the ingredients, mix thoroughly and refrigerate. Use as a spicy additive.

Makes about 1 cup

Cattail Pickles !

Harvest cattails at the height of summer, when they have large stems about 1/2 to 1 inch in diameter. Cut the cattails at water level, remove seed heads, then peel the stem to the inner white core. Wash this pith under running water. Lay out on counter or board and slice into 6-inch pieces. Place pith pieces in pint jars, add your favorite pickling recipe. (Use hot cider vinegar and 1 teaspoon pickling spices (commercial) to each 8-ounce jar of vinegar.)

Stems yield approximately 4 to 6 jars, depending on the diameter of the pith and the length of the stems.

Makes 4 pints

Cactus Pickles x

2 quarts prickly pear fruits
2 cups honey
2/3 cup cider vinegar
3 ounces red cinnamon candies Or
    1 small cinnamon stick, ground

Put cinnamon candies or stick in a cheesecloth bag. Place the fruit in pint jars. Mix the honey and vinegar in a saucepan and add spice bag or cinnamon candies to pan. Bring to a boil, then reduce heat and simmer for 5 minutes. Remove the bag and pour the liquid over the fruit to 1/4 inch from tops. Process in a hot water bath for 25 minutes.

Makes 2 pints

Pickled Mustard Flowers !

1 cup mustard flowers, washed
8 ounces cider vinegar
1 teaspoon pickling spices

Put flowers into a clean, wide-mouthed 8-ounce canning jar. In a non-corrosive saucepan, bring vinegar to a boil. Add pickling spices and simmer 'till flavors mix, about 5 minutes. Remove spice bag and pour vinegar over the flowers to within 1/2 inch of top of jar. Loosely cap the jar and place in a 2-quart sauce pan filled with water. Bring to a boil and boil for 5 minutes.

Makes 1 jar

Rose Petal Butter x

    1 cup rose petals, 2 cups
    ½ stick sweet butter

Alternate layers of petals with butter, so you get a "marble" layered appearance. Seal in a container and store in refrigerator. Blend when ready to spread on crackers or bread for a wonderful appetizer! Or, cut into little patties for another type of appetizer.

Milkweed Bud Pickles    **Cattail Pickles**

Desert Meal.

Sundries

Wild foods are fantastic vegetables and make great teas, but have you ever considered them as sundries or accessories? The following recipes give you more ways to use wild foods each day.

Nettle Hair Softener !

> 1 cup dried nettles
> 2 cups water

Simmer the nettles in the water for 20 minutes, then allow the mixture to cool. Strain.

Rinse your hair with the liquid. Let it stay in your hair for 2 to 3 minutes, then rinse with nettle liquid again. Then rinse with clear water.

> **Makes about 2 cups**

Cactus Hair Softener !

> 1 cup cactus skins or pads, chopped
> 2 cups water

Simmer the cactus for approximately 30 minutes until almost dry. Pulp will turn water thicker. Cool and rinse hair with liquid after washing. Rinse again with clear water.

> **Makes 2 cups**

Note: You may choose almost any type of pulpy cactus such as prickly pear, nopolitas, cow's tongue.

Yarrow Tooth Powder or Paste !

> 1 piece of burnt toast
> yarrow leaves, dried and ground

Scrape the charcoal off a piece of burnt toast. Make a mixture of half charcoal and half ground yarrow for a refreshing, cleansing tooth powder. Or add a small amount of water to form a paste.

Variation: Dry a thick, straight-stemmed yarrow plant. Remove and save the leaves for tea. Cut the desired stem toothbrush to size. Using a knife, cut into one end over and over to make a fuzzy tip.

In an emergency situation, cut a large raw yarrow leaf and scrub your teeth by pressing the leaf to them and rubbing. Rinse.

Wild Facial Lotion and Skin Softener !

    1 cup wild lettuce leaves, washed
    1 cup mullein leaves, washed
    1 quart water

Combine the leaves and water in a 2-quart saucepan. Cover and bring to a boil, then simmer gently for 30 minutes. Strain.

Put the resulting liquid in a bottle for use as a skin softener.
Makes about 1 quart

Wildflower Punch Bowl x/~

    1 large plastic cake dome to fit inside punch bowl
    1 cup each violets, daisies, red rose petals, multi-colored hibiscus, and pansy flowers
    several rose leaves
    Several small ferns

Invert the cake dome and add 1 inch of water. Place the violets in the water and freeze solid. Repeat the procedure with each type of flower. Use the rose leaves and ferns for the last layer.

Cover and keep frozen until ready to use. When you are ready to use it, unmold and float in a punch bowl with your favorite punch recipe.

Flower Cake.

Barbeque Cactus Burrs.

Desert Snake Stick (top).

Typical Desert Walk.

Gathering Desert Tumbleweed.

Part IV: Reference Section

Cultivation of Wild Edibles

During our Adirondack homesteading years we made many attempts at gardening, using conventional methods. Indeed, experimenting with gardens was always a summer pastime.

The 175-foot strip garden worked best. I had found out when the electric company had last defoliated the areas, beneath a power line, and since it was many years ago, I chose the strip because it was already partly cleared. The garden was high, sunny, dry, and basically rockless, as Adirondack soil goes.

I enjoyed every moment of clearing the land, especially the weeds. As I seeded with vegetable seeds, I was able to supply the family's immediate food needs, harvesting a year's supply of lamb's quarters, chickweed, sheep sorrel, violets, strawberries, and dandelions. The amount of our wild stored food continued to grow, from dried to canned to pickled. The family enjoyed nutritious meals of vegetables, spices, and herbs that were both wild and cultivated.

But after two years of fixing fences and hunting predators, of purchasing cloth, saving newspapers, and recycling plastic as covers to beat the frost, I finally turned to wild foods exclusively. Long after the cultivated vegetables were gone in fall, the wild foods continued to grow under the snow, under hay, and in cold frames. I asked myself why I had spent all that time and energy gardening when wild foods grew with such abandon, untouched by predators, insects, and weather. The answer was simple; the problem *was* getting *used* to the change in eating patterns.

Over the years we learned to use the foods nature provided so readily, and in the process discovered a whole new culinary world rich in flavor and nutrition.

A Field of Wild Foods

Gardens as we know them are basically artificial, filled with introduced species and forced arrangements. We don't all have acres of wild land near home, however, from which to forage our daily meals. On the other hand, many of us can consider changing that backyard garden or lawn into a productive field of wild foods. Weeds are more natural, and can supply much of your food needs.

Your Private Wild Food Area

I recommend stringing off a special area—perhaps a 4 by 6-foot section—of rich loamy garden soil or area previously tilled. Don't use herbicides on the soil; usually the weeds already there make great herbicide-resistant garden starters. (Only approximately one percent of all weeds are poisonous.)

For best results, transplant foraged weeds after a light rain, when young plants are more easily dislodged and the soil clings to the roots. I use a spoon for shallow-rooted types such as white clover, plantain and garden sorrel, planting them in the front rows about two to three inches apart. Experience has taught me that these plants will eventually mat and form a close-knit crop. Taller medium-rooted plants like milkweed and mustard are transplanted with a trowel and placed six to eight inches apart behind the "shorties." As these larger plants grow, they'll branch out, so to minimize encroachment on other more traditional plants, place rocks, peat moss or hardwood chips around the border of your wildfood crops. Pine needles are also excellent for this purpose. Even if you let only twenty feet of lawn return to its natural state, you have a productive food strip, or field.

This new food strip will be a natural field; eventually free of poisons.

For example, if you have been using a weed killer, the chemicals have been killing off sheep sorrel, dock, plantain, dandelion, and clovers. As wild "weeds" they will return almost immediately. At this point, harvest only the youngest leaves for minimal chemical content while your soil replenishes itself. Long-term lawn treatment requires more time. Visit a local hardware store or nursery, or call the US FDA or County Extension Service, for information on residuals for chemicals. Also, there are several books on replenishing your soil the natural organic way without pesticides.

Saltiness and High Concentrations of Nitrates

Liming the soil at the base of your plants corrects both the alkalinity and nitrates. Desert dwellers should also water more frequently since this will reduce the alkaline salt buildup.

Unless it's a rainy season, your weed garden should be watered as much as your traditional vegetable garden—about three times a week; more in arid desert areas. Keep the plants moist, but not soaked. I mist my wildfood strips with a spray attached to a hose.

Compared to plants like tomatoes, wildfoods respond twice as fast to water, fertilizer and light. And they're hearty; in a heavy frost you don't need to cover them.

Harvesting

After only a week, I'm usually able to pinch back or scissor off leaves and use them as food. In fact, cutting back a few branches or leaves generally stimulates greater growth.

In the cold Adirondack winters I allowed six inches of snow (plus hay, leaf composts or peat moss) to cover my garden. This created a hothouse effect. (Boards also can act as a winter greenhouse, especially for purslane.) In this way, I was able to harvest thyme, strawberry leaves, chicory and chickweed the whole year round.

Likewise, some common garden plants are poisonous. Especially if you have children, you'll want to eliminate or at least be aware of the dangers these plants hold. Some of these plants are listed below. See the books listed in Suggested Reading for more information.

Common Poisonous Plants

American bittersweet	Cowbane
American yew	Crown of thorns
Anemone	Daphne
Azalea	Deadly nightshade
Black locust	Delphinium
Bleeding heart	Dogbane
Bloodroot	False hellebore
Iris	Foxglove
Bouncing Bet	Four o'clocks
Buckthorn	Golden chain
Buttercup	Horse nettle
Butterfly weed	Horse chestnut
Castor bean	Horsetail
Celandine	Jack-in-the-pulpit
Choke cherry	Jack-o-lantern mushroom

Jimsonweed	Rhododendron
Lantana	Rhubarb leaves
Larkspur	Scotch broom
Lily of the valley	Skunk cabbage
Marsh marigold	Snow on the mountain
May apple	Star of Bethlehem
Mountain laurel	Tansy
Narcissus	Tomato leaves
Nightshade	Trumpet flower
Poison hemlock	Virginia creeper
Pokeweed	Water-hemlock
Prickly poppy	Wisteria
Privet	Yellow oleander

Do not let this list confuse or inhibit your foraging. The stress here is on identification. Know your own property. Other good field guides are found in the library or sold in book stores. There are too many poisons in the natural world to memorize. Therefore, (1) <u>memorize</u> your edibles, (2) <u>identify</u> the poisons nearby, and <u>be absolutely sure</u> before you try a new plant (see section on foraging, Part I, page 22).

Building a Wild Foods Garden

Wild plants will grow everywhere, in desert, valley, woods, and fields. Vacant lots and sidewalk cracks show evidence of the tenacity of common plants to live. Dandelions, plantain, yarrow, chicory and violets are common in many places. So begin by looking at your local edible landscape. View the surroundings with calm deliberation and a pencil. See the abundance already there and know the ease of creating a wild foods garden.

I cannot be specific here as to soil needs, moisture requirements, and the like. You have to consider your environment, your geographic location when working with these wild plants. The "gardening" I speak of is mostly one of transference, and *support*. You can move plants to where you want them (provided you haven't taken them from a protected area) and you can provide conditions that they like. Peat moss may help keep the surrounding weeds from inundating your food strip. Mulch helps too—one food field has 6 inches of wood chips around the main plants. Leaves, rocks, pebbles also work. Sidewalks retain a lot of <u>moisture</u>. But even simple things like cedar chips or sawdust may increase the ability of eastern soil or add to the saltiness (alkali) of desert soil. In general, it is best to use indigenous soils and topsoil.

Use a spoon, trowel, or shovel to gently release the plant's grip on the soil. See the chart on the next page for guidance on plant reactions to being moved.

In developing your wild foods garden, you may want to move plants around a bit or bring in native plants that haven't otherwise appeared. Be sure to dig up all of the roots, then set in new place and pat soil in gently. Water well.

The Inside Garden

Starting a wildfood garden in your home isn't much harder than maintaining your outdoor one. Clay pots prevent the soil from becoming rancid, and positioning them away from heating or air conditioning units helps regulate temperature. Use potting soil and organic fertilizer, and don't forget to *mist* the plants three times a week.

In my Adirondacks log cabin, I transplanted dandelions and chicory into an old-fashioned wash tub, then placed it down in the cellar where the dark, damp conditions produced an amazing abundance.

Much of your indoor gardening depends on your climate, the temperature of your house, and the hours and strength of available sunlight. I sometimes put my plants in one big twig basket and relocate them to different areas. Artificial grow-lights are occasionally necessary for certain plants like clover, which requires at least 12 to 14 hours of high light per day. When introducing artificial light, at first leave it on only a few hours.

Indoor Garden.

Transplanting Wild Edibles

Plant Name	Planting Depth	Method	Ease of Adjustment
Aloe Vera	deep	trowel	perky
Amaranth	shallow	spoon	wilts
Aster	medium	trowel	wilts
Balsam fir	deep	shovel	perky
Birch	deep	shovel	wilts
Blackberry	deep	shovel	perky
Blueberry	deep	shovel	perky
Bulrush	deep	shovel	perky
Burdock	deep	shovel	wilts
Cattail	deep	saw, shovel	perky
Chamomile	shallow	spoon	perky
Chickweed	shallow	spoon	perky
Chicory	deep	shovel	wilts
Cholla	medium deep	shovel	perky
Clover	shallow	spoon	perky
Daisy	medium	trowel	perky
Dandelion	medium	trowel	wilts
Dock	deep	shovel	wilts
Evening Primrose	medium	shovel	perky
Field thistle	medium	spoon	wilts
Fireweed	medium	trowel	wilts
Goldenrod	deep	shovel	perky
Grape	deep	shovel	wilts
Lamb's quarters	shallow	spoon	wilts
Mallow (Malva)	medium	trowel	wilts
Maple	deep	shovel	perky
Meadowsweet	deep	shovel	perky
Milk thistle	deep	shovel	perky
Milkweed	medium deep	trowel	wilts
Mint	shallow	spoon	wilts
Mullein	deep	shovel	wilts
Mustard	medium	trowel	wilts
Nettles	deep	shovel	perky
Phragmities	deep	shovel	wilts
Pine	deep	shovel	perky
Plantain	shallow	spoon	wilts
Prickly pear	deep	shovel	perky

Transplanting Wild Edibles (Continued)

Plant Name	Planting Depth	Method	Ease of Adjustment
Purslane	shallow	spoon	perky
Queen Anne's Lace	deep	shovel	wilts
Raspberry	deep	shovel	perky
Rose	deep	shovel	perky
Sheep sorrel	shallow	spoon	perky
Shepherd's Purse	shallow	spoon	perky
Sow Thistle	deep	shovel	wilts
Strawberry	shallow	spoon	perky
Sumac	deep	shovel	perky
Sunflower	medium	shovel	wilts
Thyme	shallow	spoon	perky
Tumbleweed	shallow	trowel	perky
Violets	shallow	spoon	perky
Wild lettuce	medium	trowel	wilts
Willow	deep	saw, shovel	perky
Wintergreen	shallow	spoon	perky
Wood sorrel	shallow	spoon	perky
Yarrow	medium	trowel	perky

Glendale Library Plants.
(See page 276)

Edge your wild foods garden both to keep out nonedible plants and to give it a sense of unity.

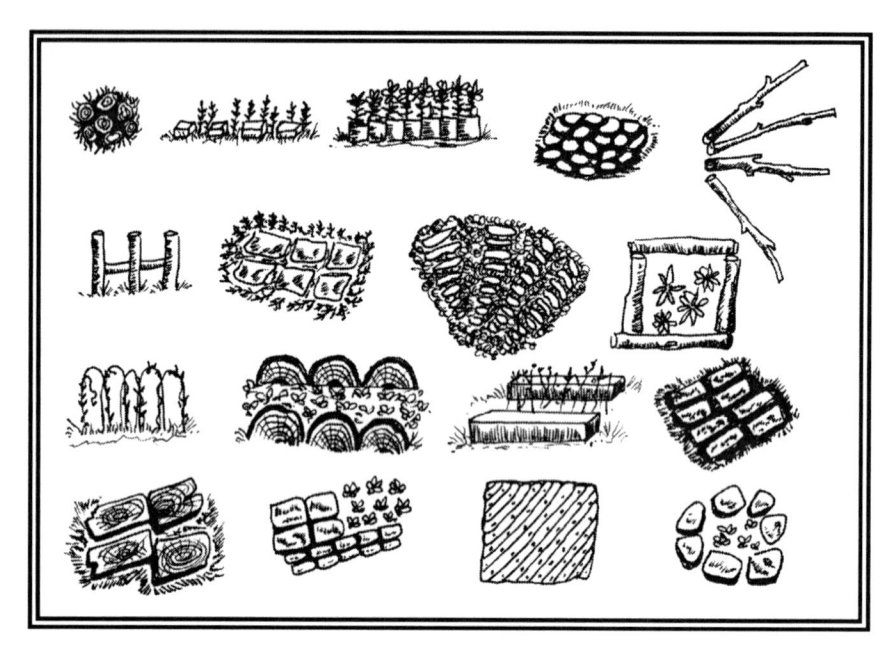

Edgings and Path Decorations

Top (left to right): crosscut circular slices of log, bricks, 2 by 4 ends buried in ground, smooth egg-shaped rocks with soil in between, straight limbs or poles.

Second row (left to right): Natural fences from wood, larger slate pieces, roundish stones, cut poles in a raised bed.

Third row (left to right): Cross fencing or picket sections, crosscut logs halved, 2 by 4's,

Bottom row, (left to right): large cut lumber boards, flagstones or bricks, decking or patio lumber, stones.

So much depended on wood.

**Illustration from *A Glimpse of Peace* and *Flickering Free*
written by Ken Heitz, illustrated by Linda Heitz (Runyon)**

**Linda's former husband Ken published two now rare books
back in the early 70's now out of print.**

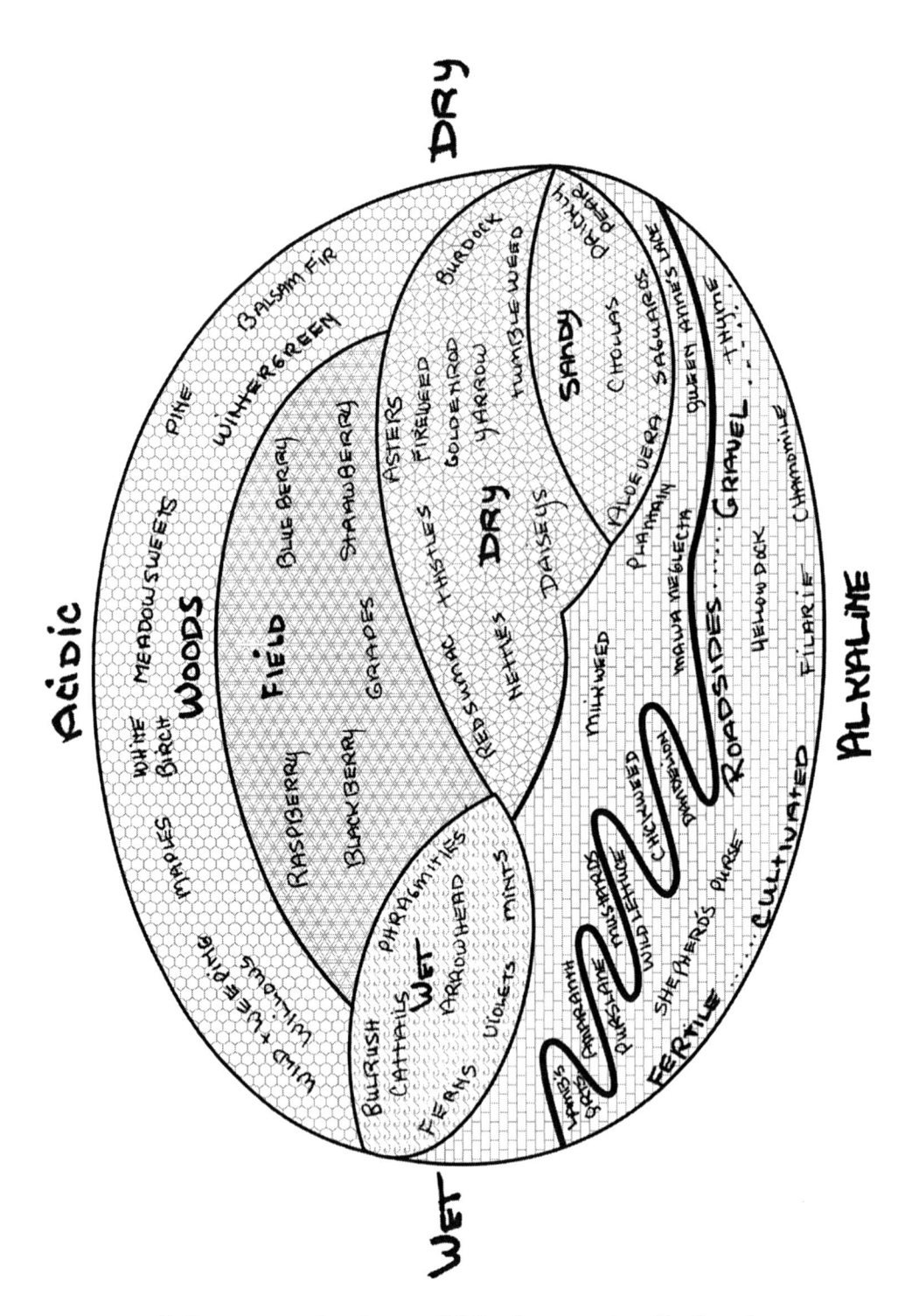

Soil-type matrix where wild foods are naturally found.

268

Legend:
A hypothetical complete Wild Foods Garden.

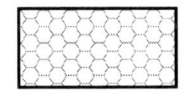

Woods. . . .Includes: wild & weeping willow, maples, white birch meadowsweet, pine, balsam, & wintergreen.

Field. . . . Includes: raspberry, blackberry, grapes, strawberry & blueberry.

Wet. . . .Includes: bulrush, , phragmities, cattails, arrowhead, ferns, violets & mints.

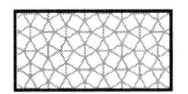

Dry. . . .Includes: red sumac, thistles, nettles, daisies, asters, fireweed, goldenrod, yarrow, tumbleweed & burdocks.

Sandy. . . .Includes: prickly pear, aloe vera, cholla & saguaro.

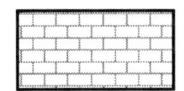

Fertile / Roadsides. . . .Includes: lamb's quarters, amaranth, purslane, mustards, wild lettuce, chickweed, dandelion, shepherd's purse, milkweed, plantain, malva neglecta, yellow dock, filarie, chamomile, Queen Anne's lace & thyme.

Left: lambs quarters   Right: sheep sorrel

Public Wild Food Identification Walk Ways

Sabael Walk 1982-1983
Sabael, New York

After her homesteading years, Linda's first home became the first public wild foods identification walk.

First year: Added topsoil, made raised beds using railroad ties. No fertilizing done. Rocked the natural areas, built bridges over swamp into woods. Peat mossed individual plants, as well as used heavy hay over winter.

Second Year: Complete regrowth of every species except sparse wintergreen patch. Presence of red dots on leaves indicated overly acidic soil

A completely natural Garden.

because of balsam trees close by (natural to area). All beds peat mossed for second summer.

> **Linda says-** I loved this first public walk around my cottage where I had electricity for the first time in 13 years. I maintained this walk for 2 years and then moved it to Rose & Bruce Burke's. This walk was studied by over 3000 people.

Burke's Cottage Garden 1983-1985
Sabael, New York

First Year: Rototilled a strip of lawn and transferred plants to rock-ringed sections. Added 6 inches of peat moss after one month's growth. Some harvesting and thinning necessary.
Second Year: Regrowth fantastic-100%.
Third Year: Good regrowth in most beds; poor regrowth of grape, meadowsweet, cattail (unattended).
Fourth Year: Plants assimilated back into mowed lawn.

Burke's Cottage Garden.

Linda says- Down the street from my home, Mr. and Mrs. Burke gave me an area to build a wild food walk. This was well marked and I gave a "tour" twice daily for two summers, before moving on to the Runyon Institute.

Note: All trees planted continued growth: birch, pine or maple. The rest returned to lawn. This walk was seen by over 3000 people.

Runyon Institute Walk
Warrensburg, New York 1986

The Runyon Institute ran for one year in Warrensburg, New York. This was the third food identification walk developed for public use. The strip was made up of common lawn plants, and the walk demonstrated the area's natural habitat.

First Year: Rototilled the strip of lawn. Added topsoil, gravel walkways, wood 2x4's in between plants transplanted from natural areas which exhibited rapid growth. Peat moss added on successive harvesting of most sections. No fertilizers used. Assimilated into lawn after first year.

Runyon Institute Walk.

Linda says- The Runyon Institute was founded by a businessman named Walter Johnson. The building was founded in my name, and it hosted over 6,000 people, showing them our wild food walk. Unfortunately, I needed to close the institute and move closer to home in North Creek where the white water rafting walks began. The Runyon Institute walk was maintained for two years. Hundreds walked the graveled paths. Trees remained and prospered.

Nantikote Lenni Lenape Walk
Bridgeton, New Jersey 1987-1988

The Nantikote Lenni Lenape developed this walk in 1987. Research done here has been invaluable for developing future food walks.

First Year: Rototilled a strip of lawn in Bridgeton City Park. Added garden topsoil. Tepee wood stays used for wheel design and plant separation. Applied 6 inches of peat moss after 2 to 3 weeks growth.
Second Year: Unattended garden prolific with growth. Largest leaf and seedling growth of all walks. Assimilated back into park after third year. I couldn't attend this one personally but once every 2 months.

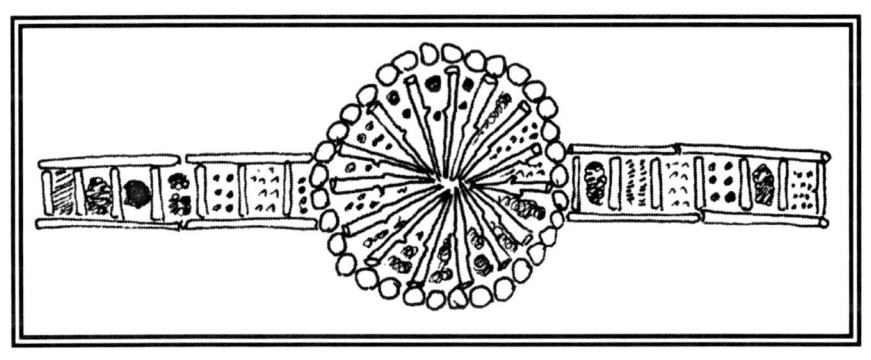

Nantikoke Lenni Lenape Walk.

> ***Linda says-*** I was asked to consider a wild food walk in Bridgeton City Park. This was Native American land, and I was pleased to work with Jim Ridgeway as well as other members of the Nantikote Lenni Lenape tribe. This walk was graciously maintained for 2 years by the Nantikote Lenni Lenapi Indian people. It closed in 1988.

"Rafting the Glenn" Walk
Warrensburg, New York 1985

First Year: Put peat moss 6 inches thick around transplanted seedlings. Natural watering. Occasionally harvested smaller "lawn foods." No fertilizers added. Some trimming done of raspberry, blackberry, roses.

Second Year: Unattended except to weed patches, add 6 inches more peat moss. All plants survived winter except for birch, pine, and cattails. Replanted more lamb's quarters.

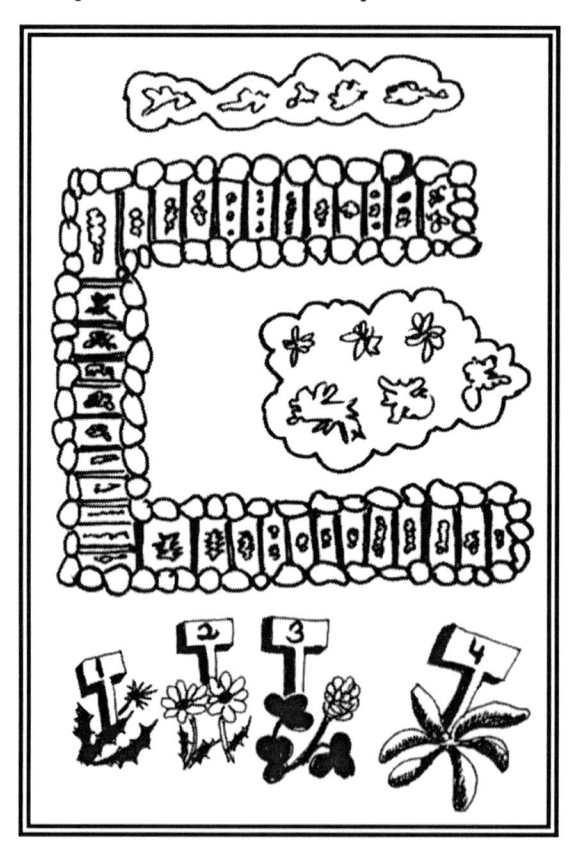

"Rafting the Glenn" Walk.

Linda says- Many of my students were young campers from white water rafting company. I particularly enjoyed the river a few yards away.

McManus Wild Food Walk
Sun City, Arizona 1990

First Year: Soil prepared to deal with caliche (desert soil). Added in equal quantities peat moss, compost, topsoil. Used raised beds, wood sides. Plants easily transplanted with spoon and trowel. Irrigated with bubbler and hose twice a month, otherwise unattended.
Second Year: Unattended; temperatures of 110 degrees for two months. Harvested remaining plants, seedlings, and seeds. Strip assimilated into neighboring areas.

McManus Walk.

Apprentice Hazel McManus.

Linda says- A gourmet cook, Hazel of Sun City, Arizona, hosted many wild food luncheons for me. Our local television stations covered these wild food functions.

Glendale Library Walk
City of Glendale, Arizona 1995-1997

David Schultz, water conservationist, was instrumental in the planning and maintenance of this walk, working along with Linda Runyon.

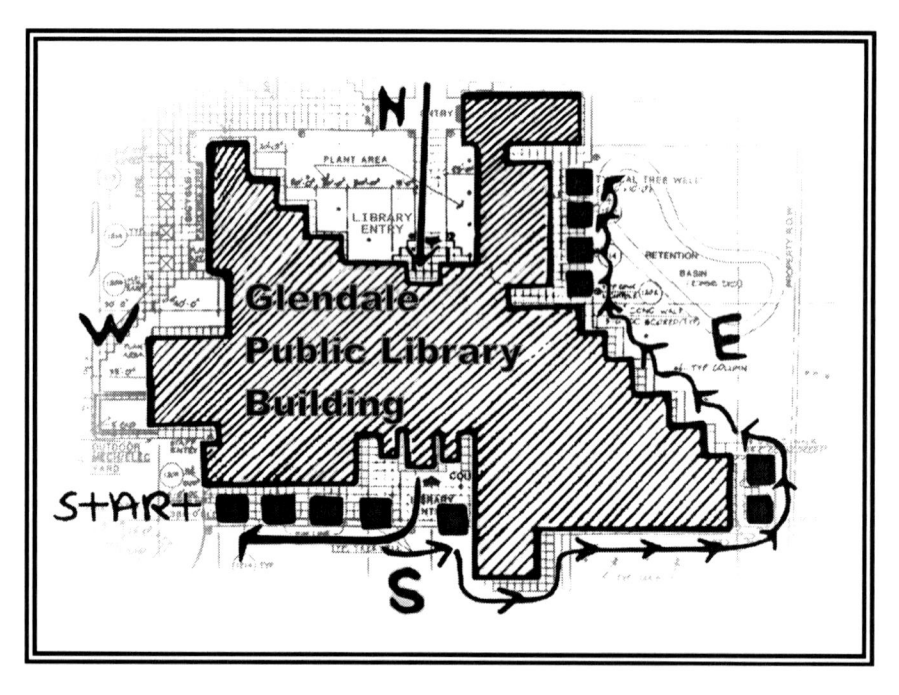

Glendale Public Library Walk.

The public visited 11 squares open in the pavement each 10' x 10', and boasting 101 wild plants and grasses in total. An underground sprinkler system was used to maintain and control this mammoth "desert" walk of wild edibles. The squares were filled with potting soil and fine desert soil and laced with peat moss. Everything was kept natural, even the use of an insect deterrent made from gigantic batches of garlic, olive oil and dish soap. Mixed in equal parts of one third each, this concoction kept the Arizona whitefly from eating the succulent new young seedlings.

Linda gives a tour.

This walk had well planned sign information using a color coded safety system: Yellow—use caution, Red—use BIG caution, and green—may eat freely after individual testing. The signs also included botanical and common names, as well as other pertinent information on edibility.

Drilling into the hard Arizona soil.
Left: John Hunkele, right: Richard Bond.

My hat is off to the City of Glendale and David Schultz. Many factors contributed to the now closed project. My hope is to bring my experience to others.

Wild Food Plant Walk
Greenwich, New Jersey, at the Children of the Earth school, started 2004

First Year: Plants were easily potted separately in native soil. The size of the pots varied to accommodate anticipated eventual growth. (Mullein grew to be 6 feet tall!) All plants were clipped and eaten from time to time, especially when a Wild Food Walk Lecture was in progress. Children kept the plants watered.

Second Year: The pots were transported to my field for the winter, and were covered in snow from time to time.

Third Year: The plants were abundant, and even more dense than the first year. Seeds fell and the soil was full of each individual plant. The Walk was used by the school for another summer. The plants wintered with the owners of the school in southern New Jersey. As of this writing (July, 2007), the plants have grown tall and quickly in the pots, just like the second year.

New Jersey Walk
Fordville, New Jersey, 2005-2006

<u>First Year:</u> This very elaborate walk required several truckloads of 4 x 4's (cut into 2 x 2's) and several loads of mulch to complete. The entire walk was 75 feet long, Because the New Jersey farm soil is so fertile, individual plants became extremely lush, spilling over in some sections and being difficult to maintain. Mullein grew to a height of 7 to 8 feet, and the Queen Anne's lace flowers were the size of cup saucers. I was by myself on this one and found it so overwhelming that I closed it after the first year. However, were there to be a crew of caretakers, such as scouts or nature park people, this type of identification center could see thousands of interested visitors.

<u>Second Year Research</u>: I went back to the walk, and saw that everything was coming back up.

1	2	3	4	5	6	7	8	9	10	11	12	13	14	15	16	17	18
										37							
36	35	34	33	32	31	30	29	28	27	26	25	24	23	22	21	20	19

- 1. White Pine
- 2. Aloe Vera
- 3. Amaranth
- 4. Burdock
- 5. Chamomile
- 6. Chickweed
- 7. Chickory
- 8. Mullein
- 9. Clover
- 10. Plantain
- 11. Dandelion
- 12. Cattail
- 13. Evening Primrose
- 14. Queen Anne's Lace
- 15. Curlydock
- 16. Wild Pepper
- 17. Sorrel
- 18. Blue Lettuce
- 19. Sassafras
- 20. Maple Tree
- 21. Nutsedge Grass
- 22. Dried Barley Grass
- 23. Crabgrass
- 24. Yarrow
- 25. Thistle
- 26. Goldenrod
- 27. Violets
- 28. Daylily
- 29. Sunflowers
- 30. Nettles
- 31. Mustard
- 32. Mints
- 33. Milkweed
- 34. Malva Neglecta
- 35. Lambs Quarters
- 36. Daisy
- 37. Willow Tree

Wild Foods Indoors

A house, apartment, roof garden, garage, or balcony are excellent areas to grow wild plants. Suitable containers for these indoor plants vary widely. Clay or plastic pots, pails, and washtubs all make ideal planters. You may even use half-gallon cans spaced evenly with a board placed on top to make an attractive shelf. There are many good books on container gardening and on growing plants indoors. Wild plants adapt easily to these ideas. Put your plants in areas of the house according to their outdoor environment. Many of them like water; the bathroom has steam, for example. Plants that like cold can thrive in a basement. I found that thyme loves heat and sun, dandelion loves the basement, near a window. Mint loves steamy areas.

The following are some plants that can be grown indoors:

Blackberry	Queen Anne's Lace
Blueberry	Raspberry
Bulrushes	Roses
Daisies	Sheep Sorrel
Lamb's quarters (western variety)	Shepherd's Purse
Mallow	Tumbleweed
Phragmities	Violets
Purslane	Yarrow

Use soil that is appropriate for the individual plants, with perhaps some rich potting soil added. Provide the suitable temperature and moisture level, and the plant will flourish. Ruth Spring, from Indian Lake, taught me to bring in dandelion, chicory, and wild lettuce for the winter. I planted them whole with their corresponding soil.

A Modest Proposal

Over 70 percent of the people in the United States live in an urban environment. More and more, urban parks are falling victim to surface pollutants, smog, ground water contamination, and, in some areas, apathy. It is harder and harder to find a clean urban area.

Potential sites for wild foods include fields, garden centers and private nurseries, environmental centers, camps, Boy or Girl Scout Centers, outdoor recreational centers and state or national parks and churches. The empty field in the middle of a city block is perfect for an urban garden. Waste areas seem to already have almost all the wild foods necessary. Add a few park benches, perhaps a water fountain, and a sandbox for the children. Plant a rototilled swatch with seedlings from nearby, edge it with wood or stones, and you have an instant identification center.

About 85 million people are added to the world population each year.[1] According to the U.S. Census Bureau, the global population doubled between 1959 and 1999, reaching 6 billion. Despite success in increasing global food production, the number of hungry people is growing. It is now thought to be 950 million—almost one fifth the world's population is underfed. Poverty is the cause. Every 24 hours, 25,000 human beings die as a result of hunger and starvation, 24 every minute, 18 of whom are children under five years of age. Acute human exposure to pesticides (such as that experienced by unprotected farm workers) is frequently fatal; 3,000-20,000 people die from pesticide poisoning annually. Human intervention to repair degraded lands takes two forms: restoration and rehabilitation. Of the two, only restoration aims to return a site to its natural state, complete with all the species that existed there before human disturbances. Rehabilitation is much more utilitarian. Its goal is to make the land productive for human use, employing whatever species and techniques which are most effective, regardless of whether they are indigenous or not. In the developing world, rehabilitation is likely to have a more prominent role than restoration because of the needs of growing population.

Perhaps the word *permaculture* is applicable here. What is permaculture? It is an attitude, a way of seeing what is before one's eyes, of enabling. It is a science, or study of natural systems such as water and plants. Permaculture focuses on *sustainable systems*, those with no pollution or waste. Permaculture is also an *ethic*. All systems of human activities support the earth's ability to sustain a diversity of life. Humans interact with their environment in mutually beneficial ways. Lastly, permaculture is a lifestyle that channels this knowledge and ethic into appropriate activities.

---

[1] World Resources, the World Resource Institute published by Oxford University, 1990.

Using wild food plants to restore the earth is the best way an individual can help at the local level. With very little help from us, the earth restores our food supply.

The urban food field is but a drawing board away. Wild foods can easily be incorporated into our educational system. Curriculum, books, tapes, videos, and newspaper articles as well as TV programs can bring this concept to the public. The environmentarian way of life is being born again!

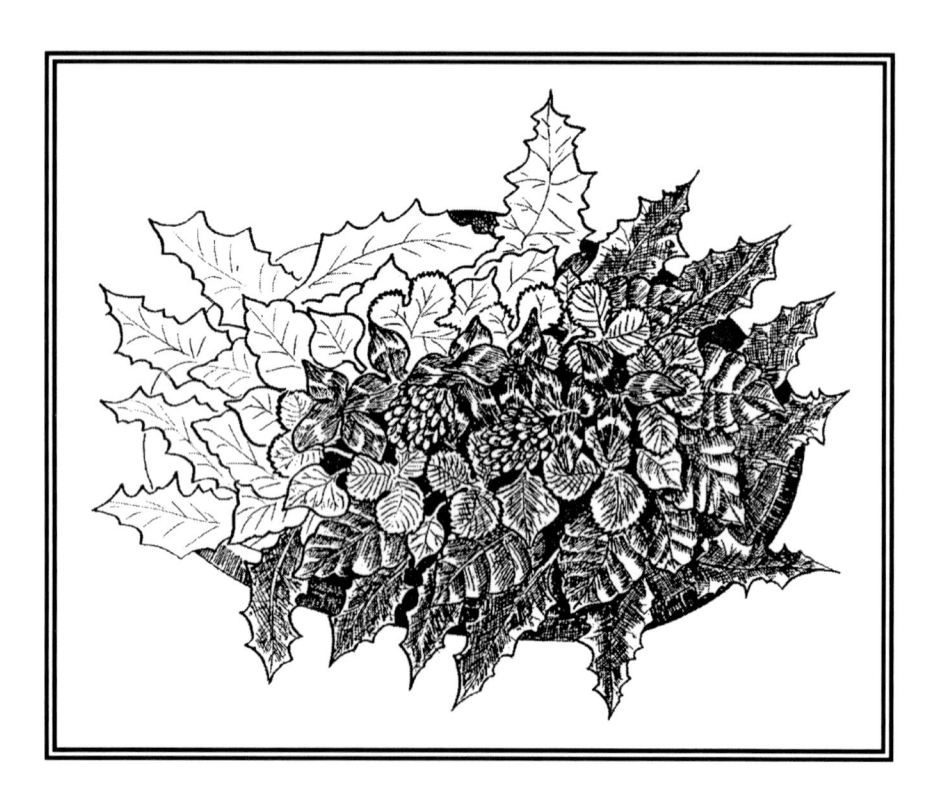

A Forager's Research Diary

There are large gaps in our knowledge of wild foods, especially in how to incorporate them into our lives. I'm always conducting experiments with these plants, seeing how well they do in a garden, whether there are new ways to use these traditional foods, and how potent the seeds remain.

The following are my notes on such experiments, along with material from the work done by my students. Growing and using wild foods remains unexplored territory.

I hope to encourage you to do your own experimentation with wild and backyard foods—not just with growing and harvesting, but in creating interesting menus for you and your family.

Purslane Experiments

Question: will dried purslane reconstitute in water? If so, will it grow and sprout roots? Can it be used reconstituted for food? Dried purslane will reconstitute well soaked in water. The plant will not sprout roots as it grows from seed. Reconstituted for food only.

August 1 — Picked a large amount of purslane. Filled a screen 1 by 3 foot. Placed screen in an inside open window receiving only afternoon sun (2:00 p.m. to 4:00 p.m.).

August 3 — Purslane roots were cut off. Black seeds from the center of the leaf clusters dropped onto the screen. There are only a few green pods in the center of the leaves still filled with seeds. Pods were apparently immature at time of picking.

September 1 — I cut off the dried, withered stem of a twig about 7 inches long and placed in water. The leaf part opened and filled with water, although the branch did not appear to grow any quicker than the one on the screen.

September 13 — Purslane in water growing well. Stems and leaves on the screen are dried and crumbly. On top of the purslane mat and up through the dried stems a completely new growth is visible. On close examination, I saw a 7-inch stem with five extensions off the main stem. Healthy green leaves extend off tips of the top three branches, with 3 tiny leaf buds appearing between the leaf extensions and joints of stem. Average size leaf is 1/2 inch. Lower two stems are weak, partially dry, flaccid at base. Small leaves extend on the tips, green and apparently healthy. All leaves are thinner than normal, stretched toward the light and do not have the succulent appearance of purslane growing in a field. Taste same. Stems are crispy and solid, about 3 1/2 to 4 inches from tip to the base. The 4 inches of base stem are apparently weak, discolored reddish, deeply veined, devoid of succulence.

<u>September 21</u> Clipped 1/4 inch off the thicker, central stem. Placed in a cup of water. Cells at the end look green and open under magnification. Stems are tightly veined and devoid of moisture. Color basically a red hue with a touch of green where the stem lay against the bottom plants on the screen. Leaves small, 1/2 inch as a norm, growing green and alert toward the western light of the window. Multiple branches tipped with green leaves have one or two small branches with small green leaves. Two necrotic twigs cut off cleanly with scissors. Cup, water, and twig placed out of direct light of window at 7:00 a.m.

<u>Conclusion:</u> Purslane dries in several days or weeks, according to moisture in the air. New growth can be seen throughout as old growth dries. Roots take in moisture from air and mat of purslane itself.

It appears that old dried purslane may be easily reconstituted in water, yielding edible food. In drought conditions, plant may be utilized with small addition of water. Purslane appears to be one of the hardiest of plants; after two months of drying, it was easily reconstituted in water.

Amaranth Experiments by Althea Dixon, student

Background: In 1992 Arizona experienced an unusually large rainfall. By May 1, we had received all the rain we usually get in a year. We had so much moisture in the adobe soil that mold was killing some of the clover and mushrooms were sprouting up. Amaranth appeared in the goat pen, thanks to these heavy rains which led to some experimentation.

August 13   I cut the first harvest from plants averaging 8 to 16 inches in height. Because of the abundant rain and rich soil, they grow with an umbrella of leaves, creating a mini-rain forest environment with lots of seedlings beneath. I remove the tops down to 4 inches above the ground.

August 17   The amaranth crop has grown from 8 to 12 inches in just four days!

August 18   Harvest two plastic grocery sacks full of tender amaranth tops. Crop is harvested to about 12" inches. Total fresh yield is 16.5 pounds.

After the harvest I notice that the soil is still well underneath, and the crop is growing without noticeable red on the stems, which is good. I can get another crop or two from this patch.

I also notice that amaranth growing in the direct sunlight does better than that growing in shade, the latter attacked by bugs causing leaf damage.

August 19   Dryers yielded 1.75 pounds of leaves and stems. After the major crop is dried, heavier stalks still have bulk and moisture enough to mold the rest of the batch. I cut stalks out with scissors and set them in an airy box for another week to finish drying. Further drying of whole batch would damage the already dry tender parts.

The high humidity is softening the tender leaves. I may have to put them in the dehydrator for an hour before whizzing them to flour. I don't know what effect this will have on nutritional value, but the color remains excellent.

August 20 Last night we had a heavy rain. I checked the progress of the amaranth and observed that undergrowth is now up to 12 inches. The day-old seedlings that were about 1/8 inch cotyledons are now between 1" and 2" seedlings, with primary leaves approximately 3/4 inches from stem to tip of the leaf.

August 24 Heavy rains over weekend. The cotyledons and primaries now have tertiary leaves and are approximately 3 inches tall. The undergrowth is now 12 inches high.

August 26 Harvested one shopping bag full. Do not weigh the green crop. Dryer yields 8 ounces of flour. By volume it makes 1 2/3 cups. Have lost only a small amount of chaff, which I sifted out of the flour as it comes from the high-impact grinder.

August 27 No rain since 8/24 in A.M. Crop has leveled off. Undergrowth continues to catch up with umbrella top at approximately 30 inches.

I harvest two more bags of tender tops this morning. These bags were the same volume as the batch on 8/18, however, they weigh only 6.5 and 5.5 pounds respectively. I fill the same number of dryer trays, but net weight of crop in dryers is 10.5 pounds. I use 1/2 pound of large stalks to make "beans."

The 4-inch seedlings are now averaging 7 to 9 inches tall with no further rain or water since 8/23. After harvesting four shopping bags of vegetable in four days, it appears there are at least 2 to 4 more harvests with no further need of water. Have not harvested red-stemmed plants; these are for seed.

August 28 Dry yield is 2 pounds 2 ounces, or 5 cups. Discarded 1 ounce of chaff plus one stem that had not dried Also lost powder in the air as it was moved from one

September 12 After running it through high-impact grinder, I get 2 quarts of flour. Dry weight is 2.13 pounds, of which 1/4 inch of heavy stem chaff is discarded.

It is interesting to note that although the heavier stems had dried hard, the humidity had gone from 20 percent to 46 percent during the wet season in Arizona, and the batch gained moisture overall. Though the texture is less crisp (the leaves would bend without breaking), I have more trouble with chibbled leaves. The color is still excellent, but more on the olive side, which I believe is due to a higher percentage of stem content in the end product. Stems dry brown whereas leaves dry green.

Conclusions:

The best procedure for harvesting to flour is as follows: Bring crop directly from garden to dryers (wash the night before with gentle water from hose) and let dry 24 hours. Rearrange trays about halfway through if dryer is old-fashioned kind with no heat or air controls. When putting crop into dryers, trim stems thicker than 1/4 inch and cook as fresh beans. This way crop is completely dry at the same time. Place directly into high-impact grinder from dryer trays; doesn't hurt if still warm. Strain chaff with a fine sieve and set to one side. Rerun the chaff all together and then sift again. What little chaff is left can be thrown away. Put flour in tightly sealed glass jars. Wrap in foil to protect from the light, and label with name of product and date. Keep in cool dry place.

Time: It takes 45 minutes to process 18 dryer trays from dried leaves to vegetable powder. It takes 30 to 50 minutes to harvest two grocery bags of fresh vegetables.

Caution: After plants form seed pods, use gloves to harvest and process flour. When processing indoors, protect lungs from flour in air by using a mask.

Reconstituted Amaranth Leaves: Can dried leaves be reconstituted to produce a palatable food item? Yes. Dried green leaves were soaked in cool distilled water for one hour, and then simmered until tender, about 5 minutes. I used only enough water to cover the leaves. The texture was so good it was almost like fresh.

Additional Comments: Was not satisfied drying amaranth outdoors. One crop got rained on and the vibrant green leaves turned sickly gray. In a survival situation you do what you can, but if domesticating amaranth, why not get the best-quality, highest nutritional product possible?

The reason for not putting amaranth in water is that if you have to leave it for any reason it can very quickly take on a dark gray color from the iron, and the quality of the dried product is not appetizing.

Further Comments: In September, 1993, the amaranth flour was inspected for color. It had remained bright green, and flavor in breads was excellent. For even longer storage, vacuum pack in plastic and store in cool dark place.

Grinding greens to flour.   **Lamb's Quarters for flour.**

Notes

Left to right: wild mustard, chamomile, dandelion and thyme.

Part V: Poisonous Look-Alikes

Here are a few troublesome plants for wild food foragers to be aware of. These are by no means the only poisonous plants out there, but they are some of the most common and are most easily confused with the plants that have been covered in the edible plants section of this book.

See also "An Introduction to Edible Plants" in Part I, especially the section on picking wild foods (page 22).

Caution: It is sometimes not the close up that is confusing, but the same plants identified at a distance can fool you.

Please Note: There are many varieties of poison hemlock, geographically spaced in all countries. Please note all species have similar characteristics. Smooth purple striated or blotched stems. Flower clusters on single stems from main stem. Hollow, circulated banded roots. Always be sure of 100% identification.
Cross reference:
Queen Anne's Lace with any white top-flowered umbrella.

Water Hemlock—Poison

Spotted Cowbane
Circuta maculata

Habitat: Wetlands, wet meadows, swamps. Canada south to Texas. This fact was observed some years ago. Do not take the state perimeter too literally.

Characteristics: The MOST DEADLY plant. Any part will kill humans. White flowers, clustered in umbrella on single stems. Stem is smooth, streaked with purple. Stem, roots chambered. (See page 292 for illustration.) All parts may even simulate smell of mild parsnip or carrot.

Caution: DEADLY POISON.

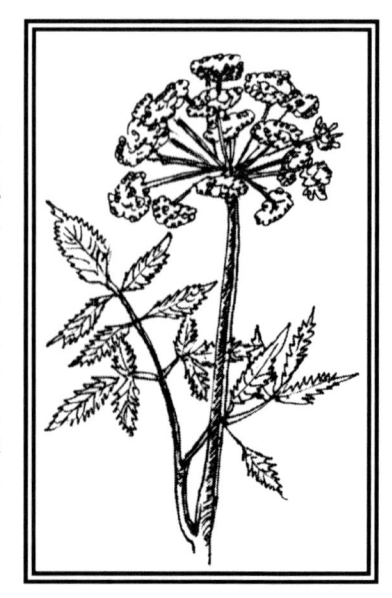

Poisonous - Water Hemlock

Poison Hemlock—Poison

Conium maculatum

Habitat: Waste ground, Iowa to Quebec South, approximately. Other genus or types on West Coast, even wet desert areas.

Characteristics: Blooms May through August. Umbrella shaped flower cluster, lacy white flowers. Dried flower maintains its umbrella shape, develops rusty colored seeds. Stalks are hairless, stout, hollow and grooved with spots of purple stripes. Small amounts may cause paralysis and death. DEADLY. Roots look like a wild carrot, but are hollow. A deadly poison. (See illustration next page.)

Poisonous - Poison Hemlock

Caution: Poison Hemlock Looks like Queen Anne's lace until flowers dry. Use 3 excellent picture references and make sure you identify Hemlock properly and DO NOT TOUCH OR INGEST. HEMLOCK IS DEADLY. (See illustration of Poison Hemlock.)

Queen Anne's Lace—Food

Daucus carota
Wild Carrot

Habitat: Waste ground fields

Characteristics: HAIRY-STEMMED biennial. Umbrella shaped flower cluster, lacy white flowers. All parts smell like a carrot. Old flower clusters "bird cage" into a seed cluster. Parsley-like leaves smell like a carrot. All stems are hairy, the root is SOLID & pithy. Smells like a carrot! (Also see page 116 in Field Guide for further description of this edible plant.)

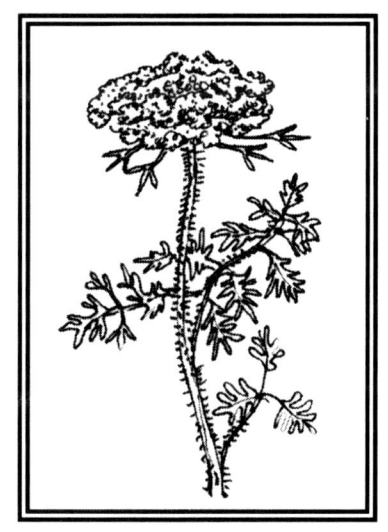

Edible - Queen Anne's Lace

(Cross Section)

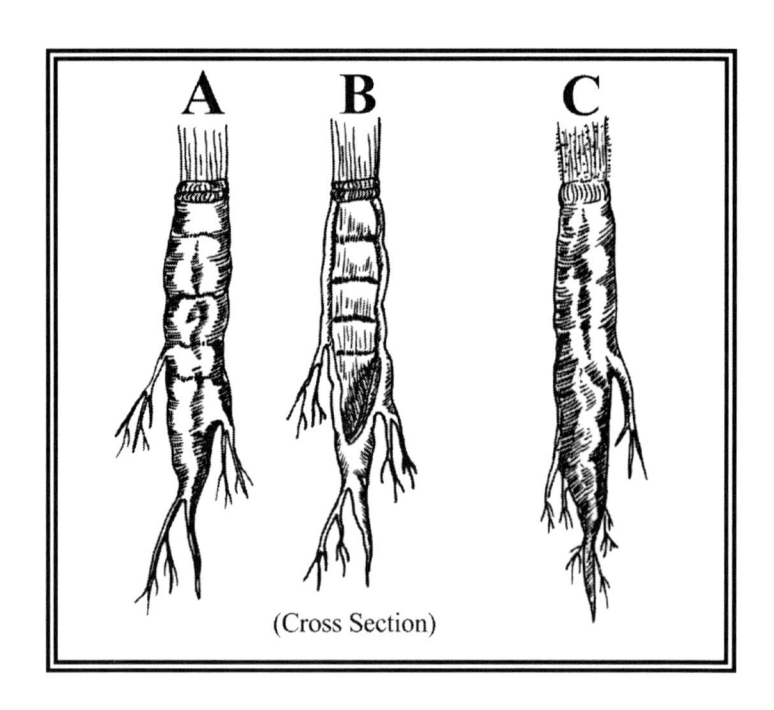

A: Poison Hemlock Root

B: Poison Hemlock root in cross section

C: Queen Anne's Lace root – food

<u>Please note</u>: Poison Hemlock (A and B) has hollow chambers with rings at the base of the stem and throughout the root. Queen Anne's Lace (C, on right) has a solid twig-like pithy root with hairy stems.

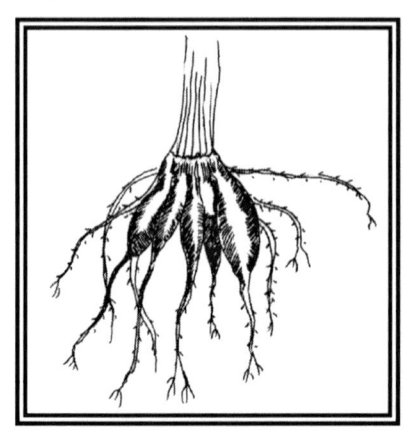

Poison - Water Hemlock Root.

Poison Sumac—Poison

Rhus vermix

Habitat: Wooded swamps.
Characteristics: Leaves, 6-12 inches long with 7-13 points, leaflets. First difference between sumac for food and poison sumac is that the poison variety has smooth leaflets. Poison berries are white. (See page 126 for description of the edible variety.)

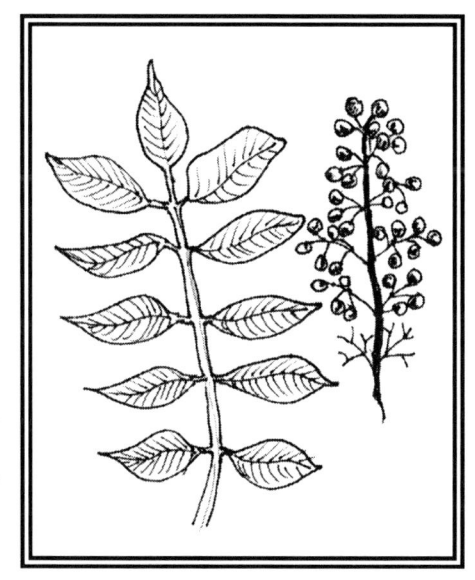

Poisonous - Poison Sumac

Edible Sumac—Food

Rhus Species

Habitat: Shrub or small tree found in all areas of the United States.
Characteristics: Twigs have milky sap. Fruit is red, hairy, and dry at maturity. Leaves: saw-tooth leaflets. (See page 126 for details.)

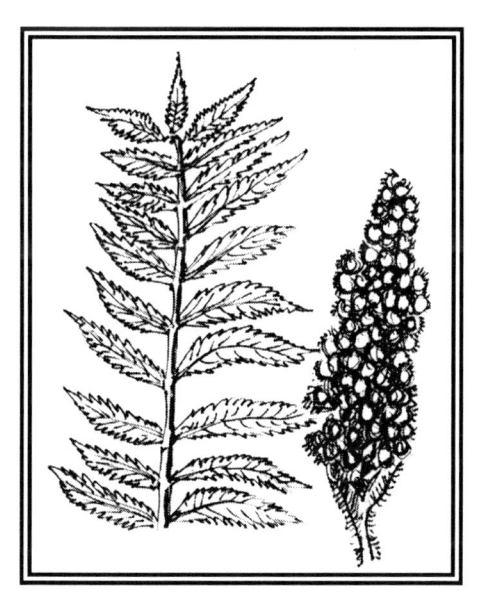

Edible - Sumac

Nettle Leaf Goosefoot—Poison

Goosefoot family: Chenopodicea

History: Naturalized from Europe.

Habitat: Moist soil, roadsides, around crops such as alfalfa, citrus, vegetables.

Characteristics: Grows to height of 1 to 3 feet, smells rank.

Caution: All raw edible parts are toxic or poison to humans.

Poisonous - Nettle Leaf Goosefoot

Lamb's Quarters—Food

Goosefoot family: Chenopodicea

History: Naturalized from Europe.

Habitat: Moist soil, roadsides, gardens, crops such as alfalfa, citrus, vegetables.

Characteristic: Grows to height of 1 to 3 feet. Smells similar to spinach. Leaves are soft and have a white powder underneath. All parts edible. Good food. (See page 93 for complete description.)

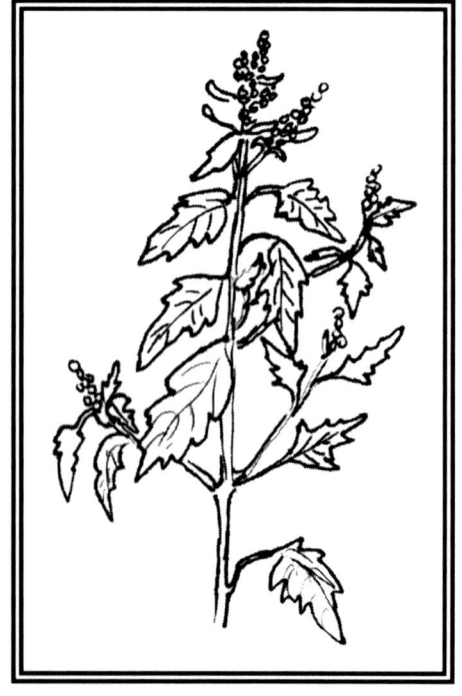

Edible - Lamb's Quarters

Field Horsetail—Poison

Equisetum Arvense

History: Family over 400 million years old.

Habitat: Wet, moist areas, swamps, lakesides, river banks.

Characteristics: Jointed green, like grass. All parts contain silicon; are toxic to humans. Abrasive silica in all parts, so plants will scrub aluminum clean.

Medicinal Value: Folk medicine wound healer, diuretic in very small amounts.

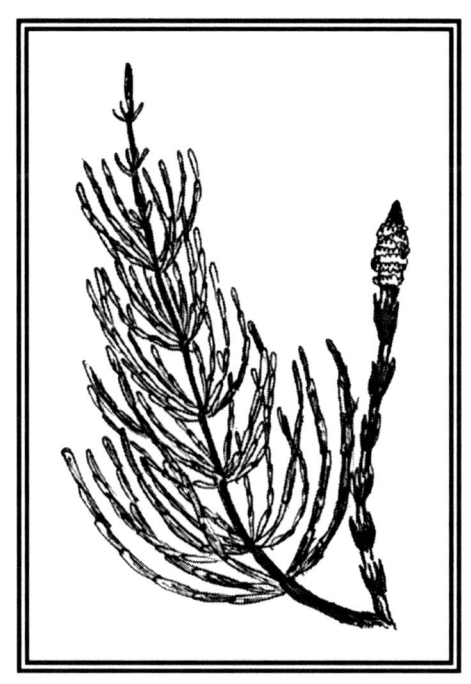

Poisonous - Field Horsetail

Grass—Food

Graminae

Characteristics: Curls and grows up like horsetail. Horsetail grows in longer grass areas. We need to examine all types of grasses carefully when choosing grass to be clipped for food. Pull out any field horsetail. (See section beginning on page 35 for complete description.)

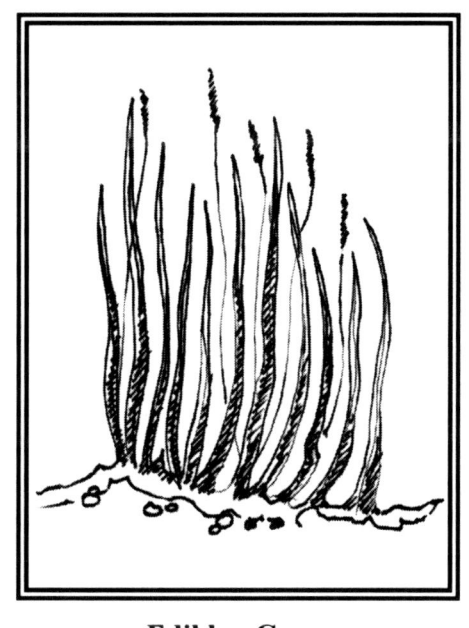

Edible - Grass

St. Johnswort—Poison

Hypericareae

Other Names:
Hypericum perforatum
History: Known for medicinal value since biblical days. Europe, herbalists.
Habitat: Fields, roadsides.
Primary Use: Leaves used as astringent and salves for bruises, wounds, anti-inflammatory.
Characteristics: Leaves shaped like lungs with many tiny holes in them. Plant grows to 1 1/2 to 3 feet. 5-petaled flowers. From a distance, plant shape is similar to the pocomoonshine variety of goldenrod.

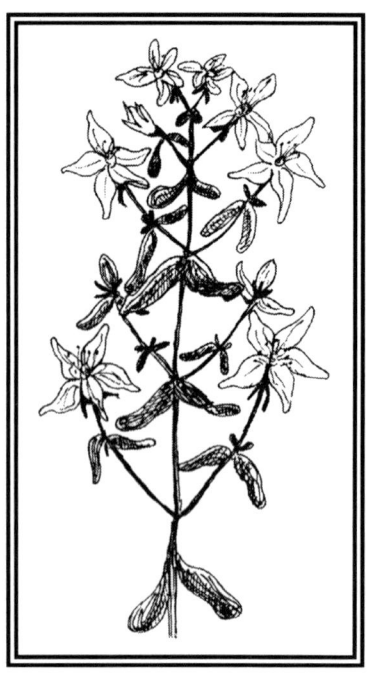

Poisonous - St. Johnswort

Pocomoonshine—Food

(Variety of Goldenrod)
Augustifolum (specific species in the *Compositae* family)

Other Names: Poco moonshine
History: Native to Europe, Asia. Naturalized in North America.
Primary Uses: Culinary, medicinal, commercial. Leaves dried for tea. Oils and extract used commercially in baking, ice cream, soft drinks, chewing gum.
Characteristics: Flowers at end of many branches 6 to 18 inches, considered an herb. The smell and uses are the same as other Canadian or sweet goldenrod species in field guide. See Goldenrod, page 90, for complete description.

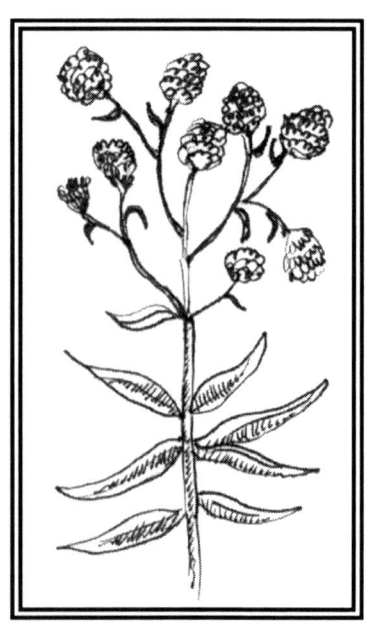

Edible - Pocomoonshine

Spreading Dogbane—Poison

Apocynum species

Habitat: Thickets, roadsides.
Characteristics: Perennial herb. Wide spreading branches, opposite leaves 2 to 3 inches long. Pink nodding bell-like flowers with pink stripe in center of each lobe. Shoots of Dogbane are hairless and quickly branching.

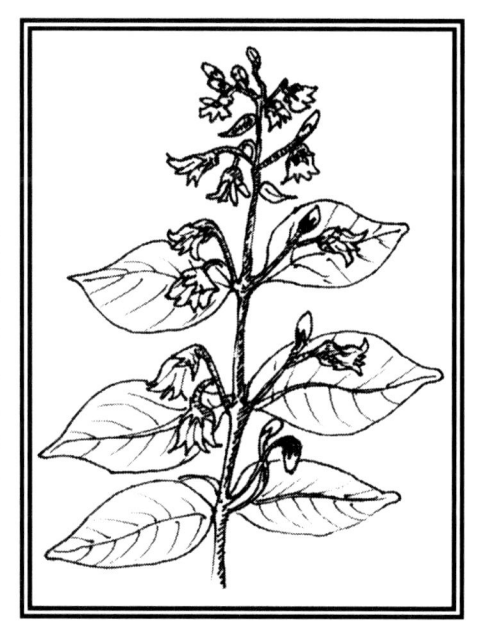

Poisonous - Spreading Dogbane

Common Milkweed—Food

Aslepias Syriaca

Habitat: Fields, roadsides.
Characteristics: Downy stemmed shoots. Leaves from 3 to 6 inches, deeply veined. Flowers dome shaped, tight, drooping clusters. After flowering, buds give way to seed pods. Crescent shaped pods, when young, may be dipped in batter and deep fried or boiled. (See page 101 for culinary uses.)

Edible - Common Milkweed

Dallis Grass—Poison

Gramineae Family

History: Native to South America, found on North American continent in the west.
Habitat: Arizona
Primary Uses: None for humans.
Caution: This grass is very susceptible to fungus attacks by paspalum ergot which is a DEADLY POISON to cattle; humans.
Characteristic: Grows to height of 1 1/2 inches. Has tight large seeds placed in a way as to attract black mold easily.

Poisonous - Dallis grass

Jungle Rice—Food

Echinochle colonum
Graminae Family

Other Names: Water grass
History: Used for food in Asia, Africa.
Habitat: Fertile desert soils of the West around crops such as sorghum, cotton, alfalfa, melons, lawns.
Characteristics: Grows 1 to 1 1/2 inches high. A sedge, but dark spots of camouflage. Areas of stems are jointed.

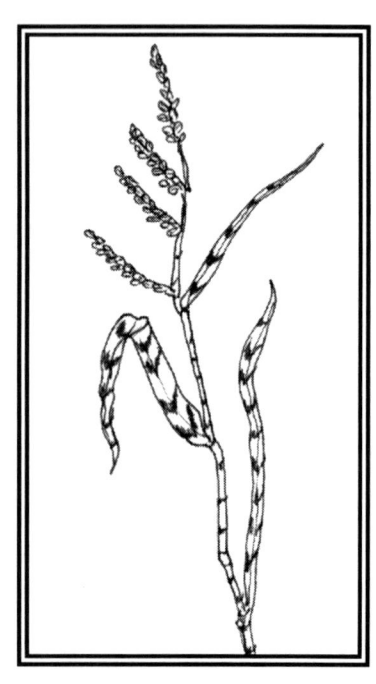

Edible - Jungle Rice

Stink Grass—Poison

Gramineae Family

Other Names: Lovegrass
History: Native to Europe.
Habitat: Found growing around base of crops such as Sorghum, Alfalfa, Citrus, gardens and waste areas.
Characteristics: Annual grows to heights of 4 inches to 2 feet. Whole plant has a cockroach like odor, rank and offensive. Flowering parts 2 to 10 inches long. Poison.

Poisonous - Stink Grass

Yellow Nutsedge—Food

Cyperus esculentus
Sedge family

Other Names: Yellow nutgrass.
History: Introduced from Europe.
Habitat: Cultivated fields, edges of crops, wet areas, ditches, roadside, streams.
Characteristics: Umbrella-like flowering tops have a thread-like projection extending from seeds and flowers. Roots have small nut-like tubers which can be used as a highly nutritious food. (See page 35 for complete description.)

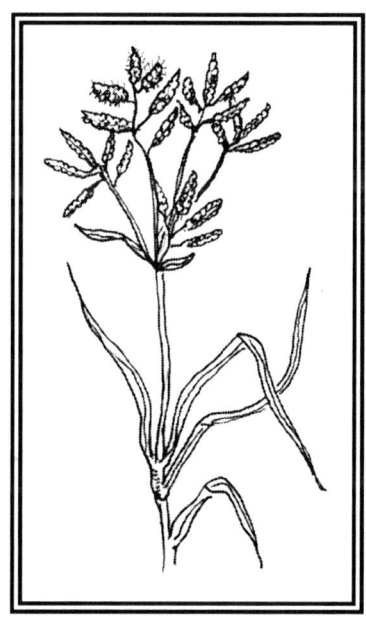

Edible - Yellow Nutsedge

Purslane—Food
Portulaca oleracea :)

Note: Purslane can be confused with White Margin Spurge, Hyssopspurge or Puncture Vine.
Habitat: Rich, sandy soils, waste places.
Characteristics: Prostrate paddle-shaped succulent leaves, tiny 5-petaled yellow flowers open in sunlight. Purslane has no milk. Leaves and stems succulent and make delicious food. Plant has a shallow tap root. (See page 115 for details.)

Edible - Purslane

Please Note: The following 3 plants are look-alikes to Purslane. When viewed from a stand height, the look of the ground cover is similar for all of these plants.

White Margin Spurge—Poison
Euphorbia albomarginata

Other Names: Rattlesnake weed
Habitat: Dry barren dirt areas along sidewalks, waste places.
Characteristics: All parts toxic, whitish milk in stem, leaves. Red blotches in each leaf.

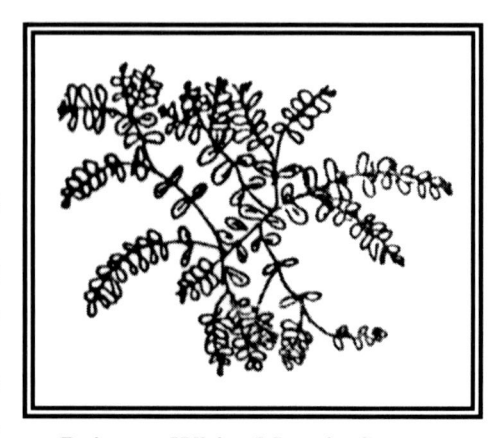

Poison - White Margin Spurge

Hyssopspurge—Poison
Euphorbia hyssopifolia

☠

Habitat: Grass, fields, crops, irrigated areas.
Characteristics: Annual. Toxic plant contains silicon in all parts. Has a white sticky milk. Grows erect, stems hairless.

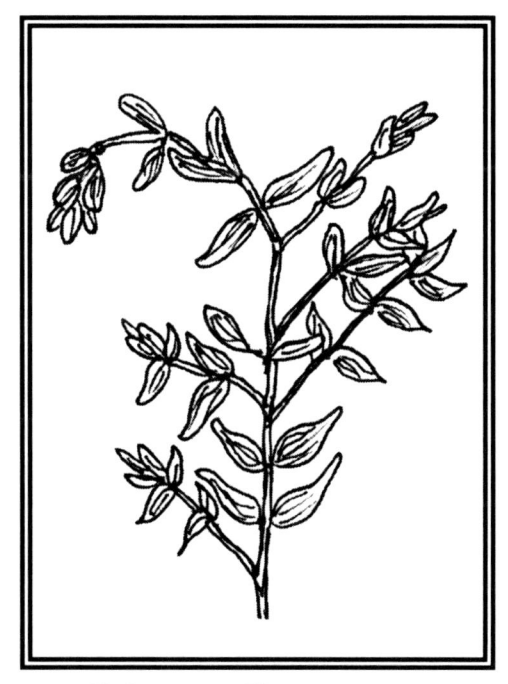

Poisonous - Hyssopspurge

Puncture Vine—Poison
Caltrop zygophyllacea

☠

Other Names: Bullhead, goathead
History: Native to Europe.
Habitat: Roadsides, gardens, fields, crops.
Characteristics: Prostrate plant has a shallow tap root, 5-petaled yellow flowers which open in sun. Seedpods contain a cluster of 5 spiny burs or nutlets that contain the seeds. Toxic to animals and humans.

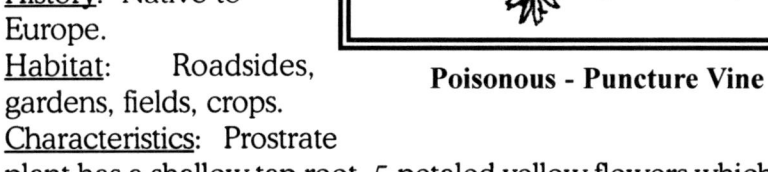

Poisonous - Puncture Vine

Appendix: Glossary of Scientific & Botanical Terms

Alkaloid: Basic, bitter-tasting organic compound containing nitrogen and forming water-soluble salts and acids.
Anodyne: Soothes or allays pain.
Antiscorbutic: Prevents scurvy.
Antiseptic: Prevents infection; kills germs.
Anti-spasmodic: Calms nervous and muscular spasm.
Aphrodisiac: Increases sexual stimulation and excitement.
Astringent: Diminishes secretion, contracts tissues.
Areole: In cacti, a clearly defined small area that may bear felt, hair, spines, glocids, flowers, or new branches.
Bacteriostatic: Inhibits the growth of bacteria.
Balsamic: Soothes mucous inflammation, especially of respiratory and urinary tracts.
Calyx: Outer whorl of floral leaves (sepals), which may be separate or fused.
Carminative: Expels gas from the stomach and intestines.
Cathartic: Purgative.
Chibbled leaves: see Compound leaf.
Compound leaf: Divided into two or more leaflets. The leaflets can be further subdivided, twice compound, or even thrice compound.
Corm: Enlarged base of a stem; bulblike but solid, not layered like an onion.
Cotyledon: First two leaves emerging from embryo.
Decoction: Liquid prepared by boiling with water and straining the cold solution.
Demulcent: Protects and soothes the mucous membranes.
Diaphoretic: Increases perspiration.
Digestive: Aids digestion.
Diuretic: Increases the flow of urine.
Dysentery: Diarrhetic condition or disease.
Elixir: Alcoholic tincture with sugar.
Emetic: Substance that causes vomiting.
Emmenagogic: Brings on or regulates menstruation.
Emollient: Soothes and softens the skin.
Essential oil: Plant oil, usually scented, that evaporates at a low temperature. Also known as a volatile oil or ethereal oil.
Expectorant: Helps expel phlegm.

Extract: Dry, soft, or liquid preparation. A fluid extract may be prepared with alcohol, distilled water, or alcohol and water.
Febrifuge: Substance that lowers a high temperature or prevents fever.
Glocids: Thin barbed bristle, produced in the aereoles of chollas, prickly pears, and a few other cacti.
Glycoside: Compound combining sugar and nonsugar units.
Hemostatic: Arrests bleeding.
Hepatic: Beneficial to the liver.
Infusion: Liquid prepared by pouring on boiling water and later straining the cooled solution.
Laxative: Mild purgative.
Leaflet: Division of a compound leaf.
Macerate: Soak in liquid to soften and dissolve.
Mucilage: Gelatinous substance swelling, but not dissolving in water.
Narcotic: Induces sleep or drowsiness.
Nerve tonic, nervine: Stimulates the nervous system.
Ointment: Salve or unction for application to the skin.
Panicle: Loose, irregularly branched inflorescence with stalked individual flowers.
Pectoral: Relieves coughing and promotes expectoration.
Plaster: Preparation spread on material and stuck to the skin.
Potherb: Plant cooked as a green.
Poultice: Soft paste, hot or cold, wrapped in cloth and applied externally.
Pubescent: Covered with fine, soft hairs.
Salve: Unctuous adhesive substance applied to wounds or sores.
Sauté': To fry quickly, stirring constantly, until lightly browned.
Seed head: Part of plant that holds a group of seeds.
Sedative: Having a calming and soothing effect.
Stomatic: Aids in digestion and stimulates the appetite.
Stimulant: Excites the functions of various organs.
Stir-fry: To fry quickly in hot oil over very high heat, stirring until food is cooked.
Styptic: Astringent, checks bleeding.
Sundries: Accessories
Tertiary: Third set of leaves of seedling; may or may not display adult attributes of plant.

Tincture: Extract of drug, usually dissolved in alcohol.
Tisane: Dilute plant infusion or tea.
Tonic: Stimulates the activity of an organ.
Tubers: Modified, swollen root.
Unction: Soothing or healing salve, ointment.
Volatile oil: See essential oil.
Winnow: To place seed heads in basket, and move up and down to allow the chaff (inedible parts of the grain) to fly out of the basket, leaving the seeds.
Whole-leaf tea: Tea made with the whole leaf, not crushed; also tea made with whole twig, not pulverized.

Edible Centerpiece

References

References in the text to Cherokee, Algonquin, and Chippewa words are based on transliterations from the *Cherokee-English Dictionary*, edited by William Pulete (Dallas: Southern Methodist University Press, 1975).

Bigfoot, Peter. *Arizona Wild Herbs.*
Densmore, Frances. *How Indians Use Wild Plants for Food, Medicine and Crafts.* New York: Dover, 1974.
Duke, James, and Alan Atchley. *CRC Handbook of Proximate Analyses.* Boca Raton: CRC Press, 1986.
Harrington, H. D. *Western Edible Wild Plants.* Albuquerque: University of New Mexico Press, 1967.
Harris, Charles. *Eat the Weeds.* New Canaan Ct.: Keats Publishing, 1973.
Hodgson, Wendy Caye. *Edible Native and Naturalized Plants of the Sonoran Desert North of Mexico.* Arizona State University: Arizona, 1982.
Page, Nancy M. and Richard E. Weaver, Jr. *Wild Plants in the City.* New York: Quadrangle, 1975.
Pedersen, Mark. *Nutritional Herbology.* Bountiful, Utah: Pedersen Publishing, 1987.
Smith, Huron. *Ethnobotany of the Ojibve Indians*, Vol. 4. Milwaukee, Wis. 1883.
Sturtevant, E. Lewis, Dr., *Sturtevant's Edible Plants of the World.* Edited by Dr. U. P. Hedrick, New York: Dover, 1972.

Suggested Reading

Agricultural Research Service. *Common Weeds of the United States.* U. S. Department of Agriculture, Washington, D.C.: Government Printing Office, 1971.

Airola, Paavo. *Are You Confused?* Phoenix: Health Plus, 1971.

Angier, Bradford. *Field Guide to Edible Wild Plants.* Harrisburg, PA, Stockpole Books: 1974

Ball, Jeff. *The Self Sufficient Suburban Garden.* Emmaus, Pa.: Rodale Press, 1983.

Birdsey, Clarence, and Eleanor G. Birdsey. *Growing Woodland Plants.* New York: Oxford University Press, 1951.

Briggs, Jim. *Wild Foods.* Hamilton County, N.Y. Cooperative Extensions publisher, 1972.

Brookes, John. *A Place in the Country.* London: Thames and Hudson, 1984.

Brown, Lauren. *Weeds in Winter.* New York: W. W. Norton & Co., 1976.

Chase, Agnes, and L. E. Bailey. *The First Book of Grasses,* Rural Textbook Series. New York: Macmillan, 1982.

Coon, Nelson. *The Dictionary of Useful Plants.* Emmaus, PA: Rodale Press, 1977.

Cooper, Hewitt Museum, the Smithsonian Institute's National Museum of Design, 712 Fifth Ave., New York, N.Y. 10019, 1979.

Crockett, James Underwood. *Greenhouse Gardening.* New York: Time-Life Books, 1989.

Densmore, Frances. *How the Indians Use Wild Plants for Food.* New York: Dover, 1974.

Duffield, Mary Rose, and Warren D. Jones. *Plants For Dry Climates.* New York: H. P. Books, 1981

Elliot, Douig. *Roots: An Underground Botany and Foragers Guide.* Old Greenwich, CT: Chatham Press, 1976.

Evers, Robert A. and Roger P. Link. *Poisonous Plants of the Midwest & Their Effects on Livestock.* Spec. Publ. #24, College of Agriculture, University of Illinois at Urbana-Champaign, 1972.

Fairbanks, Bert L. *A Principle with Promise,* Salt Lake City: Bookcraft, 1978.

Gabel, Medard. *Energy, Earth, and Everyone.* Garden City, N.Y.: Anchoe Books, 1980.

Gaertner, Erika E. *Harvest Without Planting.* Montreal: Concordia University Press, 1967.

Gibbons, Euell. *Stalking the Wild Asparagus.* New York: David McKay, 1978.

Gilbertie, Sal, with Larry Sheehan. *Home Gardening at Its Best*. New York: Macmillan, 1977.

Goldstein, Jerome. *The New Food Chain. An Organic Link Between the Farm and City*. Emmaus, PA: Rodale Press, 1973.

Hardin, James W. *Human Poisoning for Native and Cultivated Plants*. Second Edition. Durham: Duke University Press, 1974.

Harrington, H. D. *Western Edible Wild Plants*. Albuquerque: University of New Mexico Press, 1967.

Harris, Ben Charles. *Eat the Weeds*. New Canaan, CT: Keats Publishing, 1973.

Hitchcock, Susan Tyler. *Gather Ye Wild Things*. New York: Harper and Row, 1980.

Hobson, Phyllis. *Food Drying*. Charlotte, VT: Garden Way Publishing, 1983.

Kingsbury, John M. *Common Poisonous Plants*. New York State College of Agriculture and Life Sciences, New York, 1976.

Kirschenew, H. E. *Nature's Healing Grasses*. H. C. White Publications, 1960.

Lappe, Frances Moore. *World Hunger: Twelve Myths*, New York: Grove Press, 1986.

Lappe, Frances Moore. *Diet For A Small Planet*. New York: Ballantine Books, 1971.

Lerzaa, Catherine, and Michael Jacobson, eds. *Food For People, Not For Profit; a Sourcebook on the Food Crisis*. New York: Ballantine, 1975.

Luenscha, Walter Conrad Leopold. *Poisonous Plants of the United States*. New York: Macmillan, 1961.

Life Seed Foundation, Box 72, Port Townsend, WA 98368 or Native Seed Search, Tucson, Arizona

Messegue, Maurice. *Of Men and Plants*. New York: Macmillan, 1973.

Michael, Pamela. *All Good Things Around Us—A Cookbook and Guide to Wild Plants and Herbs*. New York: Holt, Rinehardt and Winston, 1980.

Miller, Saul, and Jo Anne Miller. *Food For Thought*. New York: Prentice-Hall, 1979.

Morton, Julia F. *Herbs and Spices: A Golden Nature Guide*. New York: Golden Press, 1976.

Muramoto, Naboru. *Healing Ourselves*. New York: Swan House/Avon, 1973.

New Western Garden Book. Miami: Lane Publishing, 1979.

Niethammer, Carolyn. *American Indian Food and Lore; 150 Authentic Recipes*. New York: Macmillan, 1974.

Page, Nancy M. A., and Richard E. Weaver, Jr. *Wild Plants in the City.* New York: Quadrangle/New York Times Book Co., 1975.

Peterson, Lee Allen. *A Field Guide to Edible Wild Plants; East/Central North America.* New York: Houghton Mifflin, 1977.

Poisonous Plants of the United States and Canada. Englewood Cliffs, N. J.: Prentice Hall, 1964.

Public Health Service, *Typical Poisonous Plants.* Food and Drug Administration Booklet, #017-12-00177-7.

Scully, Virginia. *Treasury of American Indian Herbs*, Prineville, OR: Bonanza Publishing, 1974.

Simmonite-Culpepper. *Herbal Remedies.* Universal-Award House, 1970.

Sheffield, Charles. *Earthwatch.* New York: Macmillan, 1981.

Soltanoff, Jack. *Natural Healing.* D. C. Warner Books, 1988.

Sourcebook on Food and Nutrition. Chicago: Marquis Academic Media, 1978.

Stokes, Donald W. *A Guide to Nature in Winter.* Boston: Little, Brown, 1976.

Sweet, Muriel. *Common Edible and Useful Plants of the West.* Happy Camp, Calif.: Naturegraph Publishers, 1962.

Szeley, Edmond Bordeaux. *The Book of Herbs.* International Biogenic Society, 1971.

University of Alaska, *Wild Edibles and Poisonous Plants of Alaska.* Cooperative Extension Service, booklet #F-40, 1953.

U. S. Department of Agriculture and U. S. Department of Health and Human Services. *Dietary Guidelines.* Washington, DC: Government Printing Office, 1980.

Waisel, Yory. *Biology of Halophytes.* New York: Academic Press, 1972.

Western Society of Weed Science. *Weeds of the West.* PO Box 963, Newark, CA 94560, Rev. 1992

Whittle, Tyler. *The Plant Hunters.* Philadelphia: Clinton Book Co., 1970.

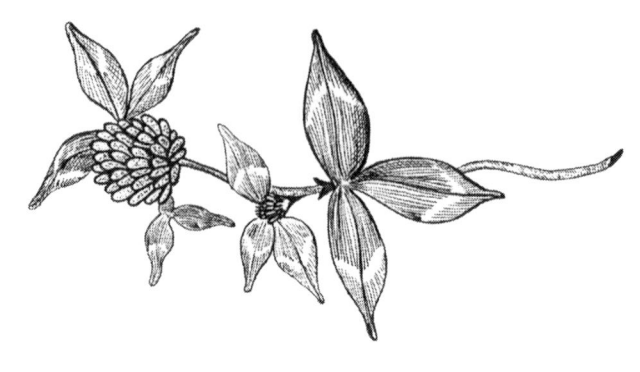

Linda Runyon